HOW
TREES
DIE

HOW
TREES
DIE

THE PAST,
PRESENT, AND
FUTURE OF
OUR FORESTS

JEFF GILLMAN

WESTHOLME
Yardley

Frontispiece: A stand of ponderosa pine, Malheur National Forest, Grant County, Oregon. (*Library of Congress*)

First Westholme paperback 2015

© 2009 Jeffrey Gillman

Westholme Publishing, LLC
904 Edgewood Road
Yardley, Pennsylvania 19067
Visit our Web site at www.westholmepublishing.com

ISBN: 978-1-59416-230-5

Also available as an eBook.

Printed in United States of America.

This book is dedicated to
Catherine and Clare
And their great-grandfathers,
Ed who liked to grow trees, and Eddie who sold their fruit.

The Wizard Tree.

It is well that you should celebrate your Arbor Day thoughtfully, for within your lifetime the nation's need of trees will become serious. We of an older generation can get along with what we have, though with growing hardship; but in your full manhood and womanhood you will want what nature once so bountifully supplied and man so thoughtlessly destroyed; and because of that want you will reproach us, not for what we have used, but for what we have wasted.

—THEODORE ROOSEVELT

CONTENTS

The author and his brother in "the Greatest Climbing Tree Ever." (*Ellen Gillman*)

Chapter One

THE TREES
AMONG US

T HE book you now hold in your hands was made from trees, but like cows raised for their meat, these trees never had a chance to reach old age. Instead, the long arm of human civilization reached out and touched them, forever changing our forests. Trees live and die at our hands. Whether felled intentionally or accidentally, most of us are aware that our trees and forests don't exist the way they once did. Exactly what has changed and how these changes have come about, however, isn't always clear.

Today's trees meet their ends after being bred, raised, and harvested like domesticated cattle. Or perhaps they perish from the onslaught of a foreign insect or disease carried by humans accidentally, or intentionally, across the ocean to our shores. Today's trees may even die from overindulgence: planted and showered with so much love and affection that they literally wilt before our eyes. Being quiet and unassuming, trees never complain about their fate. They don't scream, bleat, or whinny in protest, and this can lead us to forget that they are living and functioning beings deserving of our respect. As you will see, sometimes humans' actions are good for trees, and sometimes they're bad. My goal in this book is to show what we do to trees while they are alive, and what ultimately leads to their deaths; nothing more and nothing less. From that I hope you will draw your own conclusions about the way we treat our trees and forests.

This book about the ultimate fate of trees leads off with the story of my favorite tree. It wasn't the biggest tree, or the oldest, or the grandest, but instead, the tree whose unassuming life first made me aware of the intimate role that humans play in the lives of these plants. A tree I never knew I cared so much about until it was gone.

In a small town in Massachusetts, a little south and west of Boston and a little north and east of Rhode Island, a house sits on a lot that looks very similar to the other lots around it. But under its grassy surface lie the decaying roots of a magnificent tree. There is nothing left above ground, but in my memory this tree was the greatest climbing tree ever.

When my uncle Edward was in the fifth grade, late in 1957, a tallish whip of a tree was planted in the front yard by his father, my grandfather. Under normal circumstances the planting of this single tree wouldn't have been memorable to my uncle, or to anyone for that matter, but for the actions of my great-uncle Tom, my grandfather's older brother, whom I remember best as an old, pale, thin man with prominent purple veins on the sides of his nose. It seems that on this particular day Tom was coming over to drop off Christmas gifts. Some of these gifts were unwieldy to carry, and so Tom decided to drive over a grassy patch to park closer to the house. Tom wasn't exactly bumbling; still, he was not the sort of person that you'd want driving unsupervised through your yard. But on this particular day that is exactly what Tom did, and, upon leaving, he backed his brand-new '57 Ford Fairlane 500 right over this young tree, breaking off the top.

Throughout his life, my grandfather made a practice of caring for small things. He was the type of man who would sit like a statue for hours holding seed in his hand until birds came to feed; who would bring a nest of motherless raccoons into his house to raise and later release in the woods a few miles away. So it isn't surprising that he was upset when his tree was injured. I can only imagine the words that he must have spouted to and about Tom after the incident. (Knowing my grandfather, these rants probably lasted for years. Along with his good qualities he was known to have a bit of a temper and could hold a grudge.) Certainly my uncle Edward remembers them. They're the reason that he knows when the tree was planted. It also isn't surprising that, despite his words to the contrary, my grandfather let the tree remain in its spot instead of removing it and replanting a stronger, uninjured tree. Staking the tree and encouraging it to flourish despite its injury would have been his way. As a horticulturist I would have encouraged him to remove the tree and replant, but as his grandson I know that he couldn't.

With proper care the tree flourished, but it was a tree with many tightly clustered low branches, as a tree that has been broken off at the top tends to be if not carefully readjusted. The trunk of the tree split into three main limbs about four feet off the ground. This configuration isn't good for a tree intended for a boulevard, or for most yards, as the branches can get in the way when trying to mow the lawn, but it makes a perfect climbing tree. Growing up, my brother and I climbed this tree every chance we got, though opportunities did not often present themselves as we lived over three hundred miles away in Pennsylvania. Perhaps the infrequency of our visits causes this tree to stand out in my memory while others fade. Even today I can remember the structure of the canopy, the angle of the limbs, and even the shape of the leaves and so identify it as a maple, most probably a sugar maple. My brother and I named every branch and set of branches that we could reach: the Spiral Staircase, the Bird's Nest, the Fork, and the Main Branch, the easiest branch to use to get into the lower limbs. Part of the greatness of this tree was how easy it was to get very high. In its branches we constructed forts, played cowboys and Indians, and hid from our parents. Ricky and Tommy, my grandparents' young neighbors, would come to play when we visited, and even when we didn't. Tommy broke his arm falling from the Bird's Nest once when we weren't there. I'm a little ashamed to say that one of the first thoughts that flashed through my mind when I found out was that it served him right, climbing our tree without us.

As we got older we climbed less, drawn more to the many interesting places to see and things to do in and around Boston. I always noticed the tree when I walked by, but as I reached high school age the tree's branches no longer held the allure they once did. Then, when I was in college, I visited my grandparents and the tree was gone. The low branches, the very thing that made this a great climbing tree, had led to its demise. The branch angles were weak and narrow, and so the branches were torn off when heavy winds came through. So my grandfather had to remove the very tree that he had worked to save despite my great-uncle Tom forty years prior. My grandfather died a few years after the tree. His funeral was held on a cool, sunny April day, perfect for climbing, or for planting.

THE tree that my grandfather planted died because of the way it was treated by members of my family. Would it have fared better had it never encountered a human? Who's to say? Certainly not me. As a scientist who searches for answers about how and why trees live and die in landscapes, I often subject trees to stresses that are all but guaranteed to be terminal. I consider it my job to plant trees in ways that most landowners never would so that I can report what happens and prevent others from making the same mistakes. In the interest of research I subject trees to stresses they would rarely see in nature. Trees weren't meant to handle the depths to which I've planted them, the levels of fertilizer that I've applied to them, or the amount of herbicides that I've sprayed on them. And that barely scratches the surface of the conditions to which I've subjected these kings of the forest.

But we are all responsible for killing trees. Look through your house and note all of the objects in it that derive from wood. From the frame and floor of your home to the tables and picture frames that sit in it, there is little doubt that we rely heavily on trees as a source of structural materials and, as purchasers of these materials, it is difficult to deny that we are responsible, however indirectly, for the death of the trees from which these materials are made. Our culpability for the deaths of trees goes much deeper than simply purchasing articles made from wood, however. We have purchased land and built houses on areas that were once forests. We have cut down saplings because they have invaded our gardens or lawn. In our desire to control the environment, we have aided the movement of voracious pests from places where their populations are naturally limited by predators and parasites to places where they can grow and reproduce themselves with little to limit them. Our vanity and desire have led to the accidental importation of goods that have included inconspicuous passengers capable of devastating forests. The fact that we are unaware of the things we have done does not excuse our missteps, much less allow us to undo them.

Like a newborn child, my grandfather's tree could have had a million different futures depending upon a million different choices. My grandfather could have left it in the woods where it had been growing. He could have transplanted it to a different site. He could have chosen to end its life young because of the broken leader that the tree suffered when my great-uncle Tom backed over it. He could have given it to Tom (before he backed over it of course). He could have cut it down for kindling before I was even born. He could have pruned it properly and made it into a beautiful, if not particularly climbable, tree. Perhaps it might even be alive today if a different road had been taken. But, in the end, the choices that my grandfather made allowed the tree to live for a time in a relatively healthy though fundamentally damaged state. Ultimately the tree died because of injuries sustained through an inescapable association with humans. But then, it is also remembered because of its association with humans. Would it have lived longer or better if it had never been touched by human hands? Could it have avoided them? These are not questions with simple answers.

Chapter Two

A CHARMED LIFE

A tree can be briefly defined as a tall plant with a woody stem which survives for more than a single year. But where did they come from and why did they evolve? Plants first came onto land from the sea about 450 million years ago. These plants were small and simple, and can best be compared to algae. Slowly these organisms evolved more complexity, and about 50 million years later simple plants such as club mosses, ferns, and horsetails appeared. These plants and their relatives are still with us today. There were few animals that fed on these plants at first, and so they spread quickly across the world's landmasses, their major competition being themselves. Invading each other's space in their search for light and minerals, they struggled with one another to gain an advantage. The first plants able to grow upward and to sustain this growth over multiple years had a distinct advantage over nearby plants. Height allowed trees to intercept light before their shorter competitors, and the ability to sustain this height over multiple years meant that trees would be competing for sunlight only when they were very young.

The first trees weren't really trees as we now know them; rather, they were similar to the organisms from which they evolved. The first plant that we think of as a tree lived about 385 million years ago. These trees, now long extinct, reproduced by using spores and were more similar to a group of plants today known as horsetails than to modern trees. Closely following that group on our timeline came the genus *Archaeopteris*, a group related to ferns. Over time *Archaeopteris* and its relatives developed into the gymnosperms of today, which include such plants as pines and spruces. The nearest we can come to an ancient tree today without sifting through fossils is the ginkgo, a tree that you may well have growing in your yard, or in a nearby park. Considered a living fossil, ginkgos first appeared about 270 million years ago. These trees are most closely related to gymnosperms, though they're not typical representatives of this group by any stretch of the imagination. Plants that belong in the genus *Ginkgo* have retreated and reappeared in different regions of the world over time, depending on how favorable the climate was for their existence. Ginkgos disappeared from North America about 7 million years ago and from Europe about 3 million years ago only to reappear two hundred years ago when humans reintroduced them onto these continents from China, where they had retreated.

Most of the greenery that dominates our landscape today are flowering plants known in the scientific world as angiosperms, which are easily distinguished from the gymnosperms by the presence of flowers. Angiosperms split from their parent group, the gymnosperms, between 250 and 300 million years ago. Then, about 100 million years ago, these plants started to diversify into the progenitors of the trees we know today and to dominate the world's landscape.

So it seems that trees were here before humans set foot on earth and, unless it all ends in a big fireball, chances are good they'll be here long after we're gone. Changes in climate over the millennia have caused mass extinctions of trees and other plants as well as animals. The ancient plants I mentioned above, with the exception of *Ginkgo,* are long gone, as are most of the early angiosperms. Their descendants, however, live on, though they are often pushed from one habitat to another by continually changing climates, competition with other plants, and herbivore pressures.

During the last ice age, glaciers covered North America, forcing most plant life far south, yet today there are oaks well into Canada. About 10,000 years ago, ice sheets retreated from North America leaving fallow land behind to be colonized by whatever plants and animals happened upon it. After the glaciers receded, a variety of different plants established themselves in the newly available areas. First came grasses and other fast-growing plants and, later, forest trees. This reforestation would have taken many years to evolve. Seeds of poplars would have floated in on their silky white parachutes. Maple seeds would have swirled in on their single wings. Blue jays may have been most responsible for moving oaks rapidly northward because of the distances they can transport acorns, but mice and squirrels would have played a part too; like the blue jays they would cache their food in the fall and then forget about it at winter's end as other foods became available.

How might these trees have lived in a world with few humans? I like to think of it as something of a charmed life. These trees would have evolved within the environment they found themselves and wouldn't have been likely to encounter insects or diseases from other regions of the world. They wouldn't have been subjected to the possibility of being destroyed for today's lumber or pulp industry. But that doesn't mean that these trees would live forever. There are still plenty of dangers in the forest for a tree.

Air travel can help us understand early trees' existence in another respect. When you're in an airplane, flying level at 34,000 feet and the sky is clear, it's easy to see what has become of the huge forests that once covered the eastern portion of the United States. Down below sit postage-stamp sized fields bordered by thin lines of trees. In the spring and summer these fields are green. In the fall and winter they are tan or brown, or perhaps white. To me, the fields stand for our lost forests. For humanity to exist as we know it we need these fields to produce food and material for clothing, and yet I can't help but wonder what it might have looked like if you could go back in time, before European settlers came. The forests would have been endless, broken only by streams, lakes, meadows, and a clearing or two. In the spring and summer the forests would be green, in the winter a mixture of white, tan, and green. In the fall, depending on where you were flying, the colors would be simply magnificent. Reds and yellows with spots of green from evergreens. It's possible to get little tastes of what this flight would be like by flying over some of our national parks, or rural areas such as can still be found in northern Wisconsin or West Virginia. But, compared to what the forest must have been like once, these vistas are only a teaser of what once was.

Today, flying has become routine and is considered less dangerous than driving. Despite the sensational news coverage they receive when they occur, airplane accidents don't happen that frequently. When they do, they tend to occur at takeoff or landing. As the plane is struggling to get itself to a speed where it can

Joyce Kilmer Memorial Forest, Graham County, North Carolina. (*North Carolina Department of Tourism*)

break free of the earth's pull, there is the possibility of using too much runway or of the pilot misjudging speed. Also at takeoff, the plane is flying low enough for birds to foul the engines. At landing, there is the danger of landing gear failing or of misjudging and overshooting the runway. Compared to taking off and landing, the middle portion of the flight is quite tame.

With humans out of the picture, a tree's life is much like a flight on an airplane. At the beginning there are many dangers. Competition with nearby plants for nutrients, sunlight, and water, browsing animals, and flooding can all end the life of a young seedling. It takes a little bit of luck for the tree to survive, at first. Being dropped on fertile ground gives a seed a head start. Being on moist ground helps too, as does having good, regular rains, and a tree canopy above with enough holes so that light can filter through. But eventually the tree will reach a size where it doesn't need quite as much luck as it once did. It will be able to cope with grazing from animals. It will be tall enough to compete successfully with its neighbors for light, and it will have developed a large, strong root system to hunt down the water and nutrients it needs to grow. Certainly there are still dangers to the tree at this stage, but they seem small and insignificant compared to the dangers that it faced when it was younger. At the end, when a tree nears the conclusion of its life, as with an airplane nearing the end of its flight, there are once again a multitude of things that can bring about a tree's demise with undue speed. Older trees are less adept at producing foods for themselves than their younger siblings. Photosynthesis is not as efficient in older trees. Water makes its way more slowly through their tissues, where a lifetime of injuries have accumulated. Unseen scars beneath a tree's bark take their toll as it ages. And though the tree is no longer much threatened by grazing forest animals, other creatures dwell in its branches, slowly eroding its strength.

If a naturalist were to go back in time and document the life and growth of an oak as the glaciers retreated, I think it would look something like this:

A gray squirrel was hunting for acorns one day when he came across the grandest, most stately oak he had ever seen. Excited by his lucky find on a chilly fall day, he leapt up the strong trunk and raced toward the end of one of the many low-hanging branches to see what he could collect. First the squirrel ate. Winter was coming and he needed to insulate his small body from the coming cold. He cast aside acorns that were hollow from weevil larva, choosing instead the heavy acorns, full of meat and oil. After eating his fill, he collected acorns to cache, clambering quickly but carefully down the trunk, mindful of hawks that might find him as tasty as he found the acorns, and also careful to avoid other creatures that might try to steal his prize. He dashed through a copse of birch and into a small open area in the forest. There he buried his store, to be recovered later when food was scarce, and dashed back to the tree to collect more.

The acorn lay in the ground, in a shallow hole dug and filled by the squirrel. As winter approached, no chipmunk or squirrel or jay came to disturb its rest. The seed sat, apparently unresponsive to the moist soil around it. Inside the seed, a small embryo with all the genetic material necessary for another tree lingered in its shell. Around this embryo sat two cotyledons loaded with fats and starch, storage places for food for the young plant when it emerged and the very reason why the squirrel coveted the acorn in the first place. Both the embryo and the cotyledons were dry, the water having been pulled from them as they ripened so that, when winter came, there would be no water in the seed's cells. Water would expand as it turned into ice and would rupture the acorn's cells, quickly killing it.

As the cold of winter spread across the forest, snow covered the ground and protected the acorn from the worst that January and February had to offer. Inside the tiny embryo, chemicals were stirring and reacting, unseen by the world outside, preparing the embryo to grow into a tree. The squirrel kept himself busy and his stomach full by collecting acorns from his caches all across the forest.

Like three quarters of the acorns collected and cached by the squirrel, this acorn was forgotten in the spring rush as new foods became available, fresh foods that hadn't been left to sit through a long winter. As the snow melted, it saturated the soil surrounding the acorn until, finally, the hard outer shell softened and started to allow water into the interior. The young seed took the water up, expanding as it did so, and ruptured the now weakened shell.

The embryo, using energy from the cotyledons, first developed a strong root that expanded downward through the soil to collect the moisture and nutrients that the oak needed as it grew. But to continue to grow, the oak had to acquire more energy than just that supplied by the cotyledons; it needed a renewable way to make its own food. Soon after the root entered the ground, the growing tree sent a pair of leaves skyward. These first leaves took carbon dioxide from the air and mixed it with water from the roots. Through energy provided by sunlight, these ingredients fused to form sugar and oxygen. The oxygen was released into the air while the sugar was used to fuel the plant's growth.

Over the summer the tree continued to grow. The roots spread through nearby land largely unfettered by other trees, mining for the nutrients that the tree needed to continue on. Nitrogen, the fundamental element needed for protein, was the ingredient that most limited the tree's growth, though phosphorus, potassium, and many others were also required and captured by the tree's roots. Carbon dioxide flowed into pores on the plant's leaves and was converted by sunlight to sugars that the tree could use to grow ever taller.

But the oak was not alone. Other trees lived nearby and were growing, and even competing with the oak for resources. Birches were here, as were poplars and hazels. Because of the size of the acorn's cotyledons, the oak was able to get a jump start on its competitors. Soon, however, the other faster-growing trees caught up and overtook the young oak.

As time went on the oak grew more slowly. In the light shade of the growing forest the oak's leaves became larger and thicker, which allowed the tree to use the filtered and reflected light more efficiently. Fallen leaves from nearby trees had slowly built up on the forest floor and were eaten by fungi and bacteria in the soil, eventually becoming a part of the soil itself. In this way the forest floor replenished its nutrients, recycling all of the litter dropped by the trees. Animals helped to add nutrients to the soil too. Birds and squirrels, bears and mice ate their fill and left their wastes on the young forest floor for the trees to use.

As the glaciers retreated, a small stream had formed nearby. The level of the stream was a bit lower than the site where the oak sat and so its roots stayed moist without becoming wet. The stream also provided a small area of forest where trees wouldn't live and blot out the sunlight that the little oak needed to survive. The extra sunshine allowed the oak to grow larger than its siblings. It also helped that the stream was full of fish, which birds and bears would eat. After the animals feasted, they would release excess nutrition onto the forest, often near the little tree, helping to fuel its growth.

When winter came the little oak was again lucky. Near it had grown many brambles, which kept the animals from coming too close, but the brambles weren't so near that they smothered their little neighbor. Many plants were grazed upon by animals, but with the brambles' assistance the little oak was passed over. Other young trees were eaten down to the ground. As spring came, grazed-upon young plants either had enough energy stored in their roots to produce a new shoot or they didn't. Or, in some cases, they had enough energy to produce a new shoot but it was all for naught because of a passing animal's appetite.

Over many years the little oak turned into a larger oak. Other, smaller trees that hadn't been as lucky as our oak died from lack of sunlight, from grazing, or from too much or too little water. But our oak was in just the right spot. Its uppermost limbs stretched for the sunshine and grew at the expense of lower limbs, which had become shaded over time and were now inefficient at gathering energy from the sun. Most of the tree's limbs that were lower in the canopy were eaten by passing animals or died for some other reason, anything from collapsing under a buildup of snow to being fed upon by insects. Lower limbs did not thrive. In the upper portion of its canopy the oak produced acorns, some of which were eaten by weevils and other creatures, and some of which were collected and spread by squirrels much as the acorn that our oak came from was spread so many years ago.

On the oak's leaves a whole separate world lived. Small galls emerged on the leaves' surfaces as tiny wasps injected the leaves with chemicals that caused the leaf tissue to pucker, forming pockets where the wasp's eggs would be protected from predators as they grew. Though the growing baby wasps fed on the leaf tissue that surrounded it, not enough tissue was damaged for the tree to be truly injured. On the underside of the leaves, aphids and leafhoppers pushed their tiny straw-like mouths through the outer skin of the leaf and into its softer interior tissue, where the sap flowed, and on the outer edge of the leaves, caterpillars chewed, taking massive bites to fill their bodies with nutrients before they had to crawl into the ground to pupate and, eventually, become moths.

The oak wasn't a safe haven for these insects though. There were plenty of birds, mammals, and even other insects eating the pests that fed on the tree. These predators served as the tree's protector and kept things from getting out of hand. Robins hopped from limb to limb feeding on caterpillars wherever they could find them. Ladybeetle and lacewing larvae scoured the tree for aphids and young leafhoppers to feed on, and even the larvae of syrphid flies were getting into the act, feeding on small creatures that stole nutrients from the tree's leaves. Though the tree's leaves were eaten, there was nothing in its appearance or growth that indicated stress.

An ancient oak nearing the end of its days in Valley Forge National Historical Park. (*Jeff Gillman*)

As the uppermost leaves of our tree reached the top of the forest canopy a funny thing started to happen. Instead of increasing furiously, the growth of the top of the tree actually started to slow down. Despite the wealth of sunlight intercepting the tree's canopy, it could not put forth quite as many new shoots and leaves as it once had, and those shoots that did appear were shorter than they once were. The distance between the tree's roots and its top had become too great, and the tissues in its stem had become too clogged for this sort of extravagance. But the tree did continue to expand its limbs and over time, through luck and age, it managed to break through the canopy and become one of the dominant trees in the forest. Very little light reaches the ground in a healthy forest during the spring, summer, and early fall. The uppermost trees catch most of the sunlight and the little that filters through is caught by trees lower in the canopy and by shrubs. In the depths of the forest, there are surprisingly few plants on the forest floor because of the scant fuel remaining to drive photosynthesis.

By becoming the king of the forest, our little oak had sealed its own fate. Lightning found it, not once or twice, but many times, and its vascular tissue was shredded so badly that food and nutrients couldn't be transported up and down the trunk. Without this flow, the tree's trunk became weak. Where once only a few insects lived, thousands now teemed, feeding on the once-healthy inner bark of the tree, making what was a bad situation worse. Woodpeckers came to fish the borers out of the dying tree, finding themselves a meal, but ultimately doing worse by the oak. Diseases started to infest and weaken the wood of the tree, preparing it for its ultimate fall.

As the tree slowly lost its life, it transferred much of its energy into reproduction. A heavy crop of acorns adorned its branches. Squirrels gathered these and cached them for the winter, forgetting many and continuing the cycle of life. For three years the tree stood, not quite alive and not quite dead. Leaves would sprout periodically from various places along its trunk, but there was no mistaking this tree for a healthy tree, or even one that would recover from its condition. As holes began to emerge in its trunk, the tree became a habitat for wild creatures. A screech owl found a home in one hole, a nest of squirrels in another. Woodpeckers were starting to find the tree more enticing as an ever-increasing abundance of insects settled into the now easily penetrated bark. The tree stood like this for many years. The residents of the tree changed over time, but the population always increased. Its insides were digested by termites, bacteria, and fungi. But still the tree stood, not yet weak enough to fall and still with a role to play.

As our oak faded, other oaks began to take its place. The hole in the canopy left a spot for new oaks to grow into. The other, faster-growing trees, such as the poplars that had once shot up near our oak when it was young, were almost gone now and oaks were becoming the dominant species. The forest was strong and healthy. One day a fire would sweep through and the oaks would have to reestablish themselves, or perhaps a new disease would develop and threaten the oak's dominance, but for now the oaks were the kings of this forest.

Finally, one day, a heavy wind came. The soil around the tree had been softened by spring rains and many of its roots were failing. The wind tore the tree out of the earth, leaving its weight to rest on its nearby neighbors. Over the following years, the tree would periodically throw out a small branch or leaf, fed by the few roots that remained in the ground, but such growth was rare now. Disease and insects continued to do their work, and over time the tree returned to the soil, but not before feeding millions of forest creatures, from insects to squirrels and deer, and not before producing a few offspring of its own, many of which were lucky enough to live a life as charmed as that of their mother.

We think of trees as immortal, and oaks especially so. Even today, with all the problems that humans bring, oaks may live for hundreds of years and over many human generations. One of the slowest growing trees in the forest, the oak has wood that is amazingly strong, but ultimately even this is not enough to save it from dying. The death of a tree in a forest is not in vain, though. The tree will slowly decompose and fall to the forest floor, where it will slowly be broken into its constituent parts by the residents of the forest floor and enrich the soil, ultimately feeding other trees.

By now you probably have the sense that I consider the life of a tree a romantic thing. It's true, I do, at least the life of a tree in an undisturbed forest. But this story is just a romantic vision of the way a tree may have lived a long time ago. There have always been dangers to trees, such as floods and fires, diseases and insects, but humans and their frantic migrations across the world have placed trees on a collision course with some of the worst that nature has to offer.

Chapter Three

FORESTS
OLD AND NEW

W HEN I was seven my parents moved us from a suburb of Philadelphia to the farming community of Pughtown, Pennsylvania. There weren't that many children my age around, so until I was old enough to drive I had to entertain myself in other ways, and one of those ways was by walking to the woods nearby and looking at all of the creatures living there. It always struck me as remarkable how the little woods I lived near could contain so much life: a little stream, a tiny pond full of fairy shrimp, many different types of trees and other plants, more poison ivy than I care to remember, and of course the birds and squirrels, field mice and rabbits.

To this day when I think of a forest, this little corner of Pennsylvania (which, incidentally, means Penn's woodland) is what comes to mind. To my family, this forested area was valuable as a boundary to adjoining property and as a source of firewood for the winter, but besides these functions and as a place for me to lose myself for an afternoon, it didn't have much role in the local economy. As a youth I saw forests as relatively undisturbed places where the business of society could be forgotten, at least for a while.

Another pastime that I reveled in as a youth was reading. Books were an important part of my early life and I treasured and saved them. In fact, I still have a bookcase at my parents' house filled with books of all sorts, mostly read, but some that I still plan to get to some day. I don't recall when I learned that books were made of paper and that paper was made from trees, but I do remember when that realization struck home.

I moved to Georgia after college to earn my graduate degree by studying the insect pests of pecan trees. Pecans are grown mainly in the southern part of the state and so for my research I drove through the corridor from Atlanta in north Georgia to Tifton in the south. During this drive I was struck by the odd configuration of the pine trees, which seemed to be planted in rows. It was the most unnatural forest I had ever seen. Furthermore, though the edges of the forest were often covered with vegetation, the interior of the forest was usually devoid of the undergrowth I had come to expect. When I arrived in Tifton I asked one of the professors I was working with why the trees were in such neat rows. Upon hearing the answer I was embarrassed that I hadn't been able to deduce it myself. Those weren't forests so much as farmland, and the crop being grown was pine, some for the sawmill, and some for pulpwood, the primary raw material needed to make paper. Growing trees in this way produces an artificial-looking landscape radically different from what our forebears saw when they started to use the magnificent forests that once adorned the land.

The most pronounced feature of the Americas when explorers arrived from Europe was the size and density of the forests.

Sunlight penetrates the dense forest canopy in the Joyce Kilmer Memorial Forest. The 3840-acre memorial area is one of the few remaining examples of old growth hardwood forest in the eastern United States. The forest is home to many poplar, beech, sycamore, and oak trees, some of which exceed 20 feet in circumference, and are thought to be over 400 years old. (*Ken Thomas*)

At that time, trees were a resource second to none in importance. Lumber was needed for the construction of every large structure, from ships to houses. Wood provided fuel for stoves and cook fires, and trees and shrubs provided fruits such as cranberries and blueberries for sustenance. When explorers first came to American shores it seemed that no amount of harvesting could possibly diminish the lush forests spread out before them, but then the explorers never contemplated the eventual needs of 300 million people. As the population increased the demand for wood increased with it. In 1630, what is now the United States was roughly one-half forest land. The northern region, defined as a block with corners at Maine and Maryland in the East and Minnesota and Missouri in the West, was about two-thirds forest. Likewise, the southern portion of the United States, starting from the Atlantic and including Texas, was over 60 percent forest. The western portion was just over 50 percent forest, while the Rocky Mountain region (including Idaho, Nevada, and New Mexico, extending east up to and including the Great Plains) and Alaska made up the bulk of the nonforested region, with about 20 percent and 40 percent of these areas forested, respectively.

Today about two-thirds of our forests are classified as timberland. For land to be classified as timberland it must support enough trees to make it economically worthwhile for a person to harvest it, and not be on an area that is legally protected from harvest, such as a national park. Timberland has been an important part of our American heritage.

The East Coast of the United States has undergone a radical transformation over the past three hundred years. When settlers first landed here they quickly entered and used the forests that were available for building houses, furniture, carts, ships, firewood, and whatever else they needed. This consumption served two purposes. First, it gave the settlers the materials they needed to prosper, and second, it cleared land for farming. But the consumption also created problems. The nice thing about using old forest land for farming is a forest's soil is filled with good things, and by good things I mean dead things: old leaves, old logs, old manure from forest creatures, dead animals and insects, all clumped together on the forest floor. All of this dead material makes the floor of the forest rise over time. The dead things that accumulate are primarily composed of carbon, though there are other elements there too, like nitrogen, phosphorus, and potassium. These carbon-containing materials like to hold onto other elements instead of letting them wash away, which is what happens if a soil doesn't have enough carbon in it. Another good thing about these carbon-containing materials is that they tend to hold just the right amount of water for the roots of plants, giving young trees a useful jump start. Scientists call all of this dead stuff "organic matter."

When the settlers started to plant crops in the cleared land, they were amazed at the speed with which their crops grew. Much of this growth was due to the accumulated organic matter in the soil; however, when settlers started to plant crops on the same ground year after year, something started to happen. Trees and other perennial plants tend to hold soil together with their roots. Because of this the soil does not erode away quickly and so organic matter is able to build up. When a planting of perennials, such as trees, switches over to annual crops that are harvested year after year, organic material doesn't build up because there are no longer any dead plants in the field. When an annual crop is harvested, most of the organic material in that crop goes into people's or livestock's bellies instead of into the soil. Additionally, when annual crops are harvested, their roots cease to prevent erosion, and so the topsoil that is rich in organic matter washes away. To compound the problem, the tilling that farmers engage in to control weeds and to make the earth more inviting for crops loosens the top layer of soil, making it even more susceptible to erosion.

Despite the first settlers finding the soil of the Eastern United States to be amazingly rich, it was only a matter of time until the soil became depleted. People had to move somewhere else to find good earth, so they moved the only direction they could, west, where the soils had not yet been abused and the organic matter lost. The search for more suitable cropland is why, in the mid-1800s, so many of our ancestors moved to the Midwestern plains. Cropland in the East continued to be farmed, but with diminishing returns, and so for that reason and many others people began to build on it. Houses, roads, and shopping malls don't need the rich earth that plants do. While there were other options for the land, such as planting a perennial crop that would not take as much from the land and that could send its roots deeper into the soil, they were not usually exercised. Forest management during this time was primarily concerned with cutting up trees for lumber and firewood and pulling stumps to make room for fields. This practice was standard across the United States, but was most notable on the East Coast. By the middle of the nineteenth century, however, gold had been discovered in the western United States and people became interested in living on the West Coast, to the detriment of some of this country's oldest trees.

Redwoods were perhaps the most harshly treated trees during this stage of American history. These majestic trees are only found on the West Coast of the United States and tend to fare poorly if grown anywhere else. They are among the largest trees in the world (there are eucalyptus trees in Australia that may grow larger), reaching three hundred feet or more in height and with a trunk diameter that can surpass twenty feet. As you can imagine, the amount of lumber from even a single tree is incredible, enough to provide the framing of four respectably sized houses. The sight of these trees along the California coast was awe-inspiring, and sailors at sea often commented on their magnificence. Some seamen actually used certain redwoods as landmarks to help guide their ships to port. It probably never crossed their minds that these trees might soon be gone.

With the gold rush of the 1850s in California came a sudden need for housing. The need was first filled by trees other than the redwood. Redwoods were simply too large to handle, but Americans soon adapted their harvesting techniques in order to accommodate them. As people became familiar with the tree, the

A single redwood made up the entire train load in this 1907 photograph. (*Library of Congress*)

value of its wood became more evident. Redwoods provide a soft, easily worked lumber that is free from knots. Because of its size, it is easy to cut long, straight sections of high-quality lumber. Over the course of just a few decades, most of the California redwoods were harvested, leaving us today with only redwoods that have resprouted from stumps or germinated from seed. The redwoods that were here before the gold rush are only a memory now, but for a few. Oakland, California, once a place where redwoods ruled the forests (despite the city's name), now boasts only a single original tree. All the other redwoods in the area are young, having emerged from seeds or sprouted from stumps in the last hundred years. Called "Grandfather," this single remaining tree is over five hundred years old and sits on a steep slope. It is likely that loggers avoided cutting it because its fall would have shattered its limbs, rendering it unusable for timber. Across the rest of California there are other individual survivors of the harvest, but few areas where untouched stands remain. Such was our method of harvest at the time. Techniques have improved over the years, but historically, clear-cutting was the way to take lumber from an area. Loggers moved into an area and tore out as many trees as they could as quickly as possible. In fact, we still tend to harvest this way, but now we replant (albeit usually in oddly symmetrical rows with carefully selected trees picked to have an economic value in the future).

Unlike the lives of the first American victims of the lumber-man's saw, the lives of today's lumber and pulp trees are usually quite short. These trees are selected by professional breeders for their rapid growth, the strength of their wood and their resistance to diseases and insects. After being selected, they are planted and later harvested in huge numbers. Though these trees are used to construct our houses, furniture, and paper, we rarely stop to consider their lives, so let's rectify that by taking a look into the life of a typical pine tree bred and raised in the southern United States for the mill.

The loblolly pine was a small dot on a tray with hundreds of other small dots around it. It was a seedling, grown from the seed of a tree carefully selected for its growth rate and straight trunk, and was raised in a large plastic flat with its many siblings. A farmer bought the tree. He wanted to get out of the cotton business and plant something else, something that didn't take as much attention as the row crops he was used to. His children weren't interested in farming. They never had been, and he couldn't see any of them taking over the business. If he planted trees, the land could have value without much effort.

The tree was planted in a row along with the others in the tray. Just a few feet separated one tree from the next, and a distance just a little bit greater than the width of a tractor separated one row from another. The trees weren't tightly packed now, but they would be soon. The plot of land that the trees sat on was only 10 acres, but over the next few years the farmer would make all of his 300 acres into forestland. Most of his neighbors were doing the same. It just wasn't profitable to be a farmer anymore, and the earth here didn't grow crops like it did when he was a boy and his father had run the farm.

After he planted all the pines, the farmer came around with a water truck and roughly sprayed the ground around each of the seedling's roots. His mount was a flatbed pickup with a four-hundred-gallon water tank bolted to the back. It wasn't the best system, but it was cheap, and it worked. Most of his neighbors and the big companies in the area didn't water their trees in, counting on spring rains to do their work for them, but the farmer wanted to be sure that his trees would live so he took the extra time to irrigate. Next year he would fertilize too.

The story of industrial forests is more akin to the story of growing corn or soybeans, or even to producing cattle, than it is to what most of us think of when we think of the forests that we play or hike in. Industrial forest trees are, just like the Angus, Jersey, or Holstein cows, selected for their ability to grow and mature rapidly and provide the landowner with income. The rows of trees that I passed along those lonely stretches of Georgia road were loblolly pines, a fast-growing pine well adapted to the sandy soils of south Georgia. But if the woodsmen who planted them were serious about what they were doing, then these pines were selected even more carefully than that. Trees can be purchased based on their growth rate, form, and wood characteristics. Seeds from these trees are collected and grown to produce money-making trees that are superior to the pines that naturally grow in an area. With carefully selected trees and excellent growing conditions, it is possible to go from seedling to harvest in only fifteen years, which seems shockingly similar to the story of the cow, which once took three years to go from calf to meat and now only takes a little under two years.

The loblolly pine is native to the United States and gets its name from the locations that it prefers to grow in. Loblolly is an old English word for porridge and was adapted by settlers in the Southern United States to mean a muddy, watery location. The loblolly pine thrives in wet conditions, which is a very tough spot for most other trees. Because of this tolerance for tough spots and its quick growth, the loblolly pine is considered a great choice for a managed forest and is by far the most planted tree in the South. Currently 80 percent of all Southern pine production is loblolly pine, but there are many other trees that are grown for their wood across this country; pine is just one example. Ultimately, farmers choose their tree based on the conditions at the site, the speed with which the tree grows, and (most important) the economic demand for the tree. No one wants to grow something they can't sell, and selling is the ultimate goal of any land owner growing trees in a managed forest.

Managing forests, a practice known to foresters as silviculture, is a relatively new idea to Western society. The modern origin of forest management can be traced back to the book *Sylva, or a Discourse of Forest Trees,* which was published in 1664 in England by John Evelyn. This work is considered the first book to address the growing of trees for the production of wood. The reason that this book was developed, however, is intriguing in and of itself.

It is very easy for someone who lives in our modern society to forget how important lumber was during the beginning of the industrial age. It was necessary to have a large and ready supply of lumber to wage war, as well as for the construction of houses and as a source of energy, and nobody knew that better than the island nation of England. The English were a significant force in the sixteenth, seventeenth, and eighteenth centuries, as explorers and colonists. These endeavors required the English to defend themselves and their colonies and so they had to be able to build and sail ships. An ample supply of wood became imperative because a single ship might take as many as seventy five acres' worth of trees. The

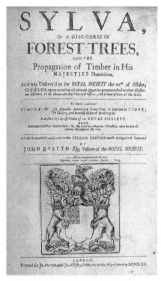

The title page of the second edition of John Eveyln's *Sylva, or a Discourse of Forest Trees*, originally published in 1664. This was the first book to describe the management of trees in a forest. (*University of Sydney Library*)

HMS *Victory*, a ship whose construction began in 1759, used over one hundred acres of oak. Forests don't appear overnight, and so, realizing this, the Royal Navy commissioned Evelyn to collect information regarding tree culture. *Sylva* was the result. This book not only outlined how to plant and grow trees, it also reviewed the more important trees growing in England at the time and, perhaps most significantly, it encouraged the growing of trees and development of forests. In later editions of the book, Evelyn claimed that his words had inspired the planting of millions of trees across England. This may or may not be an exaggeration on Evelyn's part, but *Sylva* promoted the managing of trees for the production of wood.

Growing trees in Europe was different than growing trees in the United States. The United States had a seemingly endless supply of lumber; in contrast, forests in Europe, where humans had lived for thousands of years, were scarcer and had to be managed with more restraint. There, coppicing is king, especially in England. Coppicing is a growing and harvest method that involves cutting down trees just above the level of the soil. When certain trees are harvested in this manner a new tree will usually grow from the cut stump. The greatest advantage of coppicing is that the new tree will be able to grow extremely rapidly, much faster than if it had been grown from seed. The key to the speed is the root system, which is already in the ground and can devote all its resources to a few newly emerging stems. For trees that are to become paper or particleboard, coppicing works quite well, but not all trees are intended for the paper mill and not all trees take well to coppicing.

The disadvantages to coppicing are many. Some trees, such as pine, simply won't send up new shoots. Coppicing also results in a tree that often isn't appropriate for lumber. When the new shoots emerge, they don't come out of the top of the cut stump, but instead grow from the side. Since the sun is above the plant in this situation, the new shoot will need to turn itself to grow upward. This twist isn't desirable in a piece of lumber that you want to use to build a square house. But wood from a coppiced plant wouldn't usually be used for lumber anyway. The trunks from these trees are rarely grown long enough. While coppiced trees outgrow trees that come from seeds when they're young, by the time they're twenty years old or so the seedlings catch up, and then surpass their coppiced siblings. The larger size and the straight trunks make seedlings a better choice for lumber.

Perhaps the greatest success story of selecting trees for timber and pulp comes from a tree that can be coppiced, the hybrid

poplar. Like the pine, the poplar is
a fast-growing tree that is used for
lumber and pulpwood. Unlike
the pine, poplars are considered
hardwood trees (pines are softwood). Hardwoods include such
trees as lindens, elms, maples, and
really just about any tree you can
think of with the exception of
needle evergreens like pines,
spruce, and hemlock. Many hardwoods will sprout back from coppicing to some degree, but most
needle evergreens will not.
Despite the name "hardwood,"
plenty of these trees have wood

A coppiced tree. Some trees, such as
poplars, will sprout new shoots after
they are cut down. These shoots will
grow quickly, though they do not
usually grow particularly straight.
(*Anthony D'Amato*)

that is as soft as or softer than "softwood" trees. Hardwoods are
grown preferentially in the Midwest, and the most common of
the hardwoods are the poplars. But today's poplars aren't the same
as they were when the first explorers visited.

Hybridizing, or mixing two species of trees, can be a lot of
work, but this effort can pay big dividends. Hybrids are particularly important in horticulture, agronomy, and forestry. As the
offspring resulting from crossing two different species, these trees
will inherit qualities of both parents. Although I mention crossing species casually, in practice it isn't easy, but it is possible,
depending on the plants that you want to cross. Sometimes it's as
simple as moving a tree of one species to the location of another so that they have the opportunity to mate, and sometimes it's
as difficult as taking pollen from one tree to artificially insemi-
nate the flower of another and then physically excising an immature embryo out of a seed that would otherwise fail and grow-

ing this immature embryo in a nutrient solution until it is mature enough to grow on its own.

Having qualities of two parents can be very beneficial to a plant, as one of the parents may give its offspring a trait such as resistance to a disease while the other may give the offspring a different trait such as deep roots. Regardless of specific traits, a hybrid tree tends to be greater than the sum of its parents in one very important way, hybrid vigor, which can best be described as raw, unadulterated speed. Hybrids grow extremely rapidly, and hybrid poplars are no exception to the rule. The first reports of hybrid poplar arose in the 1700s. They appear to be the result of Eastern cottonwoods from the United States crossing with black poplar from Europe and Africa. This cross produced a fast-growing tree that could be harvested just a few years after planting and then repeatedly coppiced and harvested again for firewood, with perhaps only five or six years between harvests. But the original cross is only one of many possible combinations of poplar species. There are many types of hybrid poplars, and a list of possible parents includes black poplar, big tooth aspen, cottonwood, Eastern cottonwood, and balsam poplar, along with a whole mess of others. Across the United States, universities and private forestry corporations are still working with these trees, planting them out for study and for harvest. They grow like lightning, sometimes putting on eight feet of growth or more in a year. If you're looking for a fast-growing tree to supply firewood, pulp, or particleboard, there are few trees that can match a hybrid poplar. But let's leave the poplars alone for now and get back to our little pine in its managed Southern woods. The pines of the South may not grow as fast as the poplars of the North, but they provide essentially the same things to those who farm them.

✧

The year after the pine was planted it grew more than a foot. The farmer saw that his trees were doing well and intended to keep it that way. It was late spring and the weeds were starting to take over. To keep them in check he sprayed in between his trees with glyphosate, an herbicide that kills just about anything green. He was careful not to hit his trees with the spray, a quick way to lose his investment. After the weeds were knocked down, the farmer decided that it was time to apply fertilizer. He took the leftover 10-10-10, whose name belied the fertilizer's content of nitrogen, phosphorus, and potassium, that he'd been using for years on the cotton out of the shed and applied a ton per acre of land to the trees. It was expensive, but the growth that the trees would experience was worth it. He wouldn't apply any more over the next few years, though; he didn't want to spend more money than the trees would be worth.

DESPITE Evelyn's *Sylva*, there is no cookbook for how to manage a forest. Foresters need to take into account many things, including the type of trees they're dealing with, the type of soil they have, the climate they live in, the comings and goings of humans in the area, and an incredible myriad of other things. In the southeastern United States, loblolly pine is the tree of choice. In the Northeast, spruce dominates; in the West redwoods are still popular; and in the Midwest the poplar rules. All of these trees have different needs and require different management to get them to harvest quickly. The amazingly straight rows of loblolly pines I saw in Georgia are neither normal nor odd; they are simply the way that a particular farmer, or company, decided to plant trees.

In an undisturbed forest there is a natural succession of plants and animals. Succession is the gradual change-over from one group of species to another in an area of land. First, lower-growing plants like grasses dominate, then fast-growing trees that usually have a short natural life-span move in. Finally, after all of their shorter-lived cousins fade away, longer-lived trees that have been waiting patiently on the forest floor will take over. Naturally, however, as some trees die and others fill in the gaps, the forest will change. This change is never ending. Some people refer to the presence of a "climax species" in a forest. The climax species is that species of tree that dominates a forest after a period of time because of its suitability to the conditions of that area. In reality, though, there are so many types of trees and so much that can happen to a forest, from disease to fires and windstorms, few forests maintain a single climax species for very long. In a managed forest, that isn't the case.

When people decide to manage a forest for timber, what is really being managed is succession. Pains are taken to ensure that the crop tree is grown as quickly as possible without allowing other trees or weeds that might naturally invade an area to steal water or nutrients from the crop. Whatever the cash crop is— pine, spruce, or poplar—these trees are grown to the exclusion of all else, and when they're harvested a new crop of the same thing is planted, or the cash crop is allowed to seed itself. If the area is coppiced, there is no need to plant new trees. Of course, there are different management techniques. Planting trees in nice, neat rows is only one of a long list. Foresters can also harvest a patch of land and leave a few trees of various species standing to seed the forest floor and encourage the next generation.

A clear-cut forest. Over the years the most common technique for harvesting forest trees has simply been to cut down every tree in an area, and then replant the area with the same type of tree. (*Anthony D'Amato*)

This technique doesn't give us quite the same amount of control as other techniques, but it is a system that more closely mimics what might happen in nature after a fire, tornado, hurricane, or other cataclysmic event. Foresters might also clear-cut the crop and wait to see what grows (of course they have some idea of what to expect), or they might plant a mixed crop of tree species. They could even stagger their harvest and replant trees as they harvest them so that entire areas aren't left barren. The easiest thing to do, however, is to plant, harvest, and replant, just like a field of grain, and so this is the most common technique used.

The lack of succession in a managed forest results in the loss of old growth. Old-growth trees are those that die of age and are then allowed to sit and rot. Old-growth forests support many generations of trees and are relatively undisturbed by humans. Little if any logging occurs within their confines. The loss of the old redwoods in the West is one of our most obvious and significant losses, but old-growth forests are lost almost everywhere that trees are harvested, throughout the United States and the world. When we lose these forests we lose old trees, both live and dead, including standing dead trees. Foresters have a term for these trees: snags. Snags are extremely important in the lives of certain animals. The most famous species whose habitat is threatened by the loss of old-growth forests is the spotted owl, a predatory bird that lives in the holes of dead trees in the Pacific Northwest. Another species that relies on old-growth habitat is the marbled murrelet, a seabird that lives in old-growth forests in western North America. As the number of old trees declines so does habitat for these creatures. Young trees like our southern pine, fast-growing and gone before it can become home to needy forest creatures, take the place of old trees in most managed forests, radically changing what animals can live there.

❧

Over the next ten years our pine was largely left alone. Three years after planting the farmer stopped coming to spray for weeds. Now the canopy had grown dense and so little light reached the forest floor that there was little need for him to apply herbicides any longer. Shade and dropped pine needles controlled the weeds for him. Because all of the trees came from similar seed all of the trees grew at the same rate. The trees grew straight, as any deviation to the side would mean a loss of valuable sunlight.

The farmer had planted all of his remaining acres with pine and was ready to retire. His children hadn't set foot in the forest once, and he didn't really care. If selling the land would make them happy, so be it. He had contracted with a company this year and they had gone into the pines and would harvest some to give the remaining ones room to grow. They would also apply some more fertilizer to encourage growth. The arborists had told the farmer that there were a few concerns in some nearby stands. Southern pine beetle had started to cause problems, but nothing he should worry about just yet.

ONE of the greatest problems with managed forests is that usually only one or two species are grown there, and pests tend to get out of hand fairly quickly. When you plant only one species of tree in an area to the exclusion of almost everything else then you've set up the perfect opportunity for a pest to feast. Southern pine beetles are native to the United States and like to attack trees that are living in stressful environments. These beetles live in the bark of the tree and feed on the stem's vascular tissue. This vascular tissue is the route by which water and minerals travel up the stem, and the route by which the sugars formed in the leaves travel down the stem. When this tissue is damaged, the leaves of the tree turn from a rich green to a yellowish green and then yellow and, finally, brown. The stressful environments that predispose trees to southern pine beetles are diverse and largely uncontrollable. Too much water, too little water, too much fertilizer, too little fertilizer, or too little light may all play a part in stressing a tree, as may other pests like bacteria, fungi, and insects, who look at the tree as one big smorgasbord. Crowding also causes stress, and if a field of trees isn't thinned out the trees will compete with one another, and the pine beetle will be given the opportunity to jump right in.

The usual practice for controlling southern pine beetle is to cut down all infested trees along with the surrounding fifty or so feet of uninfested trees so as to prevent a large dispersal of the beetles from the location. The good news for the person who owns the trees is that infested trees can be harvested and some money can be recovered. The bad news is that the trees probably won't get the price they would command if they stayed healthy a few more years.

One of the advantages of planting trees in nice neat rows is that they are easier to harvest when they're young. Harvest of the typical managed forest is not a onetime happening; it occurs over the course of many years. The first trees to be harvested are those that are doing poorly, don't have a good straight trunk, or are damaged somehow. These trees will be removed and sent off to the pulp mill for paper, earning the farmer a few dollars while making more room for the trees that have more redeeming characteristics. In some stands there are mixed plantings where certain trees are coppiced and others grown through seed. The coppiced plants will be harvested regularly at short time intervals for pulp while trees grown from seeds will grow larger and be harvested for lumber. Mixed-planting forests were popular in Europe before the advent of coal heat and electric power because they provided wood for fireplaces from the coppiced plants, as well as lumber for houses, ships, and furniture from the plants grown from seed. Another advantage to this sort of forest is the forest is never left barren, and old trees are allowed to stand and provide habitat for wildlife. Today, the harvest of trees from a managed forest is usually staggered, with some trees being removed earlier and others later depending on the health of the trees and the needs of the person harvesting. There usually comes a time in all forests, however, when all the trees in an area are removed and sold so that new trees can be planted.

❧

The tree was now twenty years old, a perfect size for the mill. All of its immediate neighbors had been harvested and used for pulp. The farmer had passed away in his home on the property. He left the world quietly in his sleep. The house was old and of little monetary value. The farmer's children had sold the land on which the trees stood, and now our tree was owned by a company that would harvest everything. Most trees were in excellent condition for the sawmill, but our tree had suffered some. A storm had come through and hail had badly nicked it up the previous year, knocking off some of its branches and allowing infections to take hold. The hail hadn't killed the tree, but the damage predestined its fate. This tree would become paper. It was cut off at the base and stacked with other trees. The forest would be clear-cut. Not a tree would be left standing, except near the road. There the company would leave some of the forest to minimize the visual impact of the clearing— a standard logging procedure. The trees near the road could be harvested at a later date, after the forest had been replanted with nice, neat rows. It was good public relations for the company to leave the trees near the road and make the forest seem richer in trees than it actually was.

The tree was hoisted into a truck for transport to the mill. Here the parts of the tree were separated. Bark was removed and the interior of the tree was shredded into a mixture that would be digested with various chemicals. The bark would provide energy for the pulp mill, burned to drive steam turbines. Anything left over would be sold as mulch. The pulpwood from the center of the tree would become paper.

As with our pine, trees harvested for paper are cut to pieces, supplying materials for many different purposes. Pulpwood is the inner portion of a tree. This portion is known as xylem to botanists and is the part of the tree's vascular system that transports water and nutrients from the soil up into the tree's canopy to be used in the leaves for the production of sugar and protein. Only the xylem around the circumference of the tree actually functions as real vascular tissue. The older xylem that constitutes the center of the tree, the part we use as lumber, becomes gummed up with wastes over time and loses its functionality. Old xylem serves as structural material for the tree. Xylem is made up of fibers that are very tough and durable, as you would expect since it is the primary material in wood. Besides being important for wood, toughness and durability are also what we look for in paper. The xylem fibers themselves are made up of cells that are about one-quarter inch long and one-thousandth of an inch in diameter. When pulpwood is cooked under pressure using chemicals including sodium hydroxide and sodium sulfide, the fibers separate from one another and become workable so that they can be used for paper production. These chemicals are what we smell in the air near paper mills. To make a long story short, the separated particles can then be pressed together to form sheets. These sheets constitute the paper that you are now reading from.

A paper mill is only one possible end for a tree. Saw mills may cut the tree into boards, the tree can be broken down into wood chips for mulch or particle board, or the tree may even be used for firewood. Trees are used much more efficiently today than they were in the 1800s and early 1900s when sawdust and wood chips were often thrown away (wasting as much as a third of the wood). Today, new processes are being investigated for using trees

Logs piled outside the International Paper Company Mill at Ticonderoga, New York, ready for loading by cranes. (*National Archives*)

as biofuel. The tree would be converted to ethanol using a fermentation process and then used in our cars. Besides the wood itself, there are also uses for the tree's bark. Tree bark may be used for mulch, or for the media that plants are grown in. Composted pine bark is a popular component of the soil mixes we use to grow other plants including annuals, perennials, and even trees. The pine bark can even be burned for energy, which is what happens in many paper mills. In fact, paper mills often produce more energy than they use.

In 1630, about half of North America was forest. Today, only about a third is, and most losses have occurred in the Northern region. Roughly half of the land here that was forest is now something else, a city or farmland. In the South, too, roughly half of our forestland has been converted to something else. On the West Coast, in the Rocky Mountain region, and in Alaska there have also been reductions, though not as significant.

The good news is that recently there has been a steady increase in the amount of forestland in the United States, though much of this is managed forestland. The process is slow, but from 1997 to 2002 the total amount of forest increased from 747 to 749 million acres (a three-tenths of a percent increase). Forests take a long time to grow and we can't expect miracles overnight. Unfortunately, the increase in forestland gives a somewhat inaccurate picture of the number of trees growing in the United States. Our forests are in decline. In 2009, a long-term study by Phillip van Mantgem and his colleagues showed that across the United States, especially in the West, the density of our undisturbed forests is declining. Furthermore, this decline is not just among old trees. Trees of all ages are dying, and more quickly than they once did. The greatest culprit seems to be climate change. Long periods of drought or warmth place stresses on trees, making them more susceptible to pests such as pine beetles. Even in managed forests we cannot grow as many trees on an acre of forestland as we once could. Soil that was forestland has been denuded by annual crops, and now when trees are reintroduced they struggle in the poor soil.

If our timberlands are properly managed, we will have this resource for years to come, but if we harvest trees the same way that we harvest corn, soybeans, and wheat then we will eventually deplete the ground in the same way that annual crops deplete land over many years. With our tree harvest, we remove all the resources the trees have stored over their lifetime in one fell swoop. By continually planting and harvesting the same piece of earth we doom our forests to a slow and lingering death. By crowding the same species of tree on a plot of land we promote the outbreak of pests who specialize in eating that species of tree. The United States still has some of the most magnificent forests in the world, but only through careful management will we be able to continue to enjoy them.

Chapter Four

STOWAWAY PLANTS

THE trees that we plant and harvest in our managed forests today are, by and large, native to the United States. But in our landscapes and croplands we rarely focused on these plants. Over the years we have planted more and more trees from different areas of the world, and these introduced plants have had some interesting interactions with our native species.

We live in a global ecosystem. Thanks to humans, plants can move easily from place to place, but it's not just plants that benefit from our transit. Insects, reptiles, mammals, fungi, bacteria, and even nonliving things like diamonds and oil are moved around the world with speed never before seen. Our travels have introduced new species to new places, sometimes hospitable, sometimes not. We rarely hear about the species that flounder, but the ones that flourish make headlines.

Most of the plants introduced to North America from foreign soils were not brought in by explorers, as they were in Europe, where presenting a new flower or fruit to a king or bishop was considered a noble tribute. In fact, until relatively recently, the United States was one of the more interesting places being searched for new plants, not the sender of search parties to go find them. New plants generally found their way here because of colonists who wanted to make sure they had their favorite foods or who wanted to maintain some of the more attractive plants of the old world. It wasn't until late in the nineteenth century that Americans really started searching for ornamental crops from foreign lands.

One of the best known plant species introduced to the Americas is the dandelion. Originally brought to the United States from Europe in the seventeenth century as a pot herb for medicine and wine, this flower quickly discovered that it loved American soils, and, to the chagrin of lawn owners everywhere, it has spread across most of the United States, which is somewhat ironic since most turf grasses themselves are not natives. Crabgrass, another lawn pest and also a native of Europe, was introduced in 1849 by the U.S. Patent Office as a forage crop which could also be harvested for grain if need be (ever considered making bread from crabgrass?). Kudzu, a native of Asia, was introduced in the early 1900s as a plant that could help control soil erosion. All of these crops were brought in with the best intentions, but once established they refused to stay confined and spread, making a general nuisance of themselves. Some, like the dandelion and crabgrass, are simply unsightly. Kudzu is more pernicious; it invades the edges of our southern forests, shading them heavily as it climbs trees and crawls along the ground at a rate of a foot a day, changing the mixture of species in those areas as it grows. But these are just a few of the unexpected effects that plants introduced with the best of intentions can have. Other plants have been transported to the United States solely for their attractive appearance, and these ornamentals have had some of the most devastating impact on native trees.

In any native American forest there is the opportunity for species from other regions of the world to grow and potentially flourish. If introduced plants grow faster and are able to resist or avoid attacks from local forest pests, then they may well outcompete native forest species. In managed forests introduced trees such as Scots pine and Norway spruce were once popular. Though these trees can still be seen in forests across the country, they have not grown to dominate our native trees. Today, we still plant introduced trees in our forests, though by and large the number is very small. Hybrid poplars, some of which include genes from the nonnative European black poplar, are probably the most planted semi-nonnative forest trees in this country. But introduced plants have also moved into our nonmanaged forests. Japanese honeysuckle, buckthorn, and tree of heaven, all trees prized primarily for their ornamental value, have become important forest species despite the fact they didn't originate on this continent. These plants grow quickly and can push aside their native counterparts, relegating them to small roles on the forest floor.

One of the most interesting cases of trees introduced for their aesthetic value is the Norway maple, a tree most likely introduced around 1760 by William Hamilton, a great American horticulturist of the eighteenth century. He imported this tree for his formal British gardens, which were some of the most splendid in what was to become the United States. The Norway maple made a fine addition to his grounds, fast growing and with a beautiful, rounded crown. It filled out its portion of the garden nicely and provided wonderful shade for lounging or reading. In the ensuing 250 years the tree has established itself in forests and landscapes across the Eastern United States, but it's not always the most pleasant neighbor.

The Norway maple is an aggressive tree introduced into the United States for its ability to grow rapidly and produce shade. Despite these assets this tree is considered a pest because it limits the area available for native species, such as sugar maple, to grow. (*Home Landscape Materials*)

The Norway maple enjoys the climate in the northeastern United States and, more important, it produces lots of seed that tend to fare well when they fall on the fertile soil of the forest. This maple is very shade tolerant, more so than many of our native trees, and so it is able to flourish despite the shade created by trees higher in the canopy. Once this tree starts growing, native trees can't keep up with it. The sugar maple, a native species that looks a lot like the Norway and takes up a similar place in our forests, is often placed in direct competition for growing space with the Norway. Unfortunately, the sugar maple is ill prepared for the battle. Brian Kloeppel, from the University of Georgia, and Marc Adams, from Penn State, compared the growing abilities of Norway and sugar maple and determined that the Norway maple grew faster, perhaps even twice as fast, than the sugar maple and is more efficient with its use of light, nutrients, and water. And so, even when Norway maples are removed from a forest, the seeds tend to repopulate more quickly than their native counterparts. But that isn't the entire story.

Forest services in the Eastern and Central United States consider the Norway maple an invasive plant and so want to limit its spread, or even eradicate it from their states, but control isn't easy. Sales of this maple are highly restricted. In fact, it's often illegal to sell or plant it. Foresters have tried various techniques to control it, but using herbicides is difficult because of the locations of the trees, and cutting them down creates space for more invasive plants, often including the Norway maple itself. Scientists have searched for insects or diseases from other parts of the world that could attack and retard the Norway maple's growth, but there are some major drawbacks to this method. Red, silver, and sugar maples are so similar to the Norway maple that it's entirely possible that anything brought in to feed on or infest the Norway would backfire and affect the native maples to a greater extent than the Norway maple does. And so it seems that we are stuck with the Norway maple, at least for now.

This maple doesn't stand alone as an aggressive introduced tree though; far from it. Many other introduced trees and shrubs have found comfortable niches in U.S. forests. Amur maple, buckthorn, barberry, Russian olive, and many others are taking over little bits of American forests in climates where they thrive. These plants crowd out native plants and change the profile of the forest, providing a constant challenge to foresters who want to preserve native forest ecosystems.

After people introduced them, some nonnatives spread of their own volition, but it's only fair to note that humans have helped and encouraged others. When we plant a field of plums,

we select against native crops. There are fields in the United
States that have been under cultivation for over two hundred
years. If we are growing wheat or corn in these fields then we
have effectively stopped native plants from growing on this land
for two centuries. Is the suppression of native plants through the
cultivation of exotics any worse than the suppression of native
plants by exotics that have adopted this country as their home
and now grow in the wild?

If you are a typical North American, few of the horticultural
crops you find on your table originated in North America.
Leading native crops include blueberries, cranberries, squash, and
about half the genes of your strawberries. Maple syrup comes
from the sugar maple which is, as we have seen, a native tree and
constitutes the first sweetener in the United States used by
American Indians before Europeans came. You could add tobac-
co though it's (obviously) not a food. The pecan comes from the
United States, but not the hazelnut (at least not the ones you're
used to eating) or the English walnut (though the black walnut
comes from North America). Most fruits that North Americans
favor are from elsewhere. All of the citrus fruits came from the
Far East. Some types of crabapples are native to North America,
though they bear little resemblance to the apples we usually eat.
We have many native plants related to the peach, plum, apricot,
and cherry, though their fruits aren't what you and I picture
when thinking of a fruit plate. Additionally, a number of plants,
such as tomatoes and potatoes, originate in South America and
were first introduced to Europe before they made their way to
North America.

The introduction of these and other plants to the United
States changed our ecosystem drastically and gave some insects,
plants, fungi, bacteria, and other organisms the opportunity to

flourish. Their growth in fields that were once forests altered portions of the land so that native plants were reduced to after-thoughts. Of course it can also be said that some native species were desired, and even allowed to expand their range. The blue-berry is a native of the United States that was encouraged by the European settlers. Certain trees used for timber are also highly desirable and so allowed to flourish. Oak, maple, pine, poplar, and others were and still are valuable in the United States.

When Europeans settled the New World they brought with them crops only remotely related to plants already living in the United States. If we look at all of the cropland in the United States, 99 percent of it is planted with crops that are not native, including potatoes, wheat, apples, and peaches. Europeans, as a group, can be thought of as plant collectors. Most of Western Europe—Spain, Portugal, Italy, England, France, and Germany—had explorers who traveled the globe in search of riches for their country, but these weren't necessarily gold or silver, or diamonds or rubies. The most valuable commodity explorers could discov-er was plants. These were often food plants, but they also includ-ed spices, drugs, medicinal plants, and even plants that were treas-ured simply for their beauty. Some examples of introduced plants used for food in Europe include peach, plum, pear, cherry, corn, potato, tomato, peanut, and wheat. While a diamond or gold has some value in and of itself, a diamond is just a diamond, and no matter how hard we try to make it otherwise that diamond will only have market value. Besides, a diamond cannot spawn more diamonds. But a plant is different. A plant can reproduce itself. In fact, a plant can change a whole culture if it's the right plant. And a culture can change the plant as well.

Thomas Jefferson wrote, "The greatest service which can be rendered any country is to add an useful plant to its culture." And through his words we can see how plants and their introduction into new areas were regarded until a century or so ago. When the U.S. Department of Agriculture was founded in 1862 one of its primary objectives was to introduce foreign plants to America. It is no stretch to say that, until the beginning of the twentieth century, plants were recklessly moved throughout the world with little thought as to how they would affect ecosystems. Plants such as potato, tomato, and peanut were moved from place to place as Western culture discovered and incorporated them into their diet. This movement seemed like a good idea, but it also caused problems. None of the aforementioned crops is from the United States or Europe. These crops originate in South America and were moved, first to Europe, and later to North America, for the purpose of feeding growing populations. Though it isn't a tree, the potato provides a wonderful cautionary tale about what can happen when we shift plants from one region to another without thinking ahead.

Most of us consider the potato a native crop, but it isn't unless something from South America is native. The potatoes we use in the United States originated in Peru and were introduced by explorers into Spain and other countries in the sixteenth century. In South American markets, there is a huge array of different types of potatoes, but only a few of these types ever found their way to Europe and later to the United States. Europeans first used the rather ugly vegetables to feed livestock and sometimes to brew alcohol, eschewing any larger role in a human diet because of their appearance. It wasn't until the 1620s that the potato found its way back to what would become the United States, but it didn't become an important source of food here until a steady stream of Irish cultivated its popularity through the 1800s.

The potato, after a few decades of relative neglect in Europe, eventually found a home in Ireland in the late 1700s. As a crop that flourishes in poor growing conditions, the potato just had to hang on until a nation with the right conditions found itself in need of a staple crop. Ireland during the 1700s was an impoverished country with a huge peasantry and very few resources. The potato was useful because it could be grown at a high density and provide a relatively large amount of nutritious food to a farmer and his family. In fact, though not considered optimal, many in Ireland lived on a diet of potatoes and milk and fared reasonably well up until the 1840s. Called the Irish apple, this vegetable is almost solely responsible for a doubling of the population of Ireland between 1780 and 1840. On average, it is estimated that the people of this country used about eight pounds of potatoes per person per day during that time.

The potato grew in popularity over the 1700s and into the 1800s to such an extent that Ireland's entire economy depended on it. But at what cost to the plants and ultimately the people? The potatoes that the Irish were using came from "seed" potatoes. That is not to say potatoes were grown from seed; rather, those potatoes not used for food were divided up and planted, and from these pieces grew new potato plants. If you wanted to, you could do the same with the potatoes in your pantry. Just cut off a hunk of the potato that includes an eye, plant it, and you've got a new potato plant. The problem with this sort of potato growing was that all of the potatoes in Ireland were very similar to each other. The potato offspring would have one parent instead of two and so parent and offspring were genetically identical. If one potato was susceptible to a particular disease they all were. Thus, while Ireland was awash in potatoes, their bath was a homogenous one, ripe and ready for some disease to foul the bathwater for all.

Despite all the benefits the potato brought the Irish, when we
think of Ireland and potatoes we
tend to think not of the good, but
of the bad. The great potato famine
of the 1840s destroyed a culture
and left a population starving. In
the early 1800s, reports of a disease
affecting potatoes began to emerge.
Some people paid attention, and a
few even warned that if something
wasn't done the country could find
itself on the verge of starvation. No
action was taken and so, when the
disease now known as late blight, a
malady originating in Mexico,
reared its ugly head, the potato crop
in Ireland failed—spectacularly. The
famine first struck in 1845, destroy-

"After the Eviction," 1848. The
great potato famine in Ireland during the 1840s killed or forced the
emigration of more than 2 million
people. Many of those who
remained were destitute, and some
sought shelter in ditches and
hedgerows. (*Illustrated London News*)

ing most of the Irish potato crop and by 1846 the entire crop
was lost. Between 1845 and 1860 over a million lives were lost
and over a million and a half people emigrated, many to the
United States. The population of Ireland at the time was about 8
million. Think of that: Over a quarter of the population of a
country suddenly gone because of a plant disease. The famine
happened because of humans' desire to transport plants to locations where they were never intended to be and then culturing
them to the exclusion of almost everything else. But human culpability goes deeper than that. Not only did the Irish grow these
plants to the exclusion of almost everything else, they grew only
a few types of these plants, reproducing them asexually instead of

through seed. Thus, when late blight struck, all of the potatoes were extremely susceptible to it. No potatoes were resistant. A plant that once supported a nation now crushed a nation through negligent human handling.

Meanwhile, in the United States potatoes were becoming more popular for a variety of reasons, not the least of which was that Irish immigrants, despite what had befallen the potato in Ireland, continued to cultivate it when they arrived. Potatoes were originally grown in the Eastern United States, as were most crops delivered from Europe. They spread across the country through the 1700s and were strongly established by the mid-1800s, with greater and greater populations occurring throughout the West as those regions were developed for farmland. Then a little beetle appeared on the scene.

As the potato spread west, a scary thing happened. Think about the insects in and around your home. Some have easily recognizable names. You can identify a bumblebee, an ant, or a wasp, but there are also lots of insects that just don't seem to do very much, and you end up classifying them under the generic category of bug or fly. You tend not to think about them too much. There was a beetle in Colorado that met these criteria. It was a nothing sort of beetle, with some pretty stripes, but that's about it. It ate a few plants, most of which would be considered weeds. The beetle had not even received a lasting name. Then, around the 1820s, the potato came to Colorado. Everything was fine, for a while. Potatoes grew, and this small beetle ate the plants it enjoyed, most of which farmers cared little about, and everyone was happy—until the mid-1860s. It seems this beetle had been fooling people into believing that it was just a harmless sort of insect. No one knows why the Colorado potato beetle waited so long to attack potatoes. It is thought that the insect needed to develop a taste for the plants, but once it had that taste it devastated potato crops across the western plains. In fact, this beetle was able to jump from potato field to potato field so that today, instead of being locked up in just one tiny region of the country, it eats potatoes across most of the United States. Fortunately, Americans never depended on the potato as the sole source of sustenance. Unfortunately, those who had invested heavily in potatoes lost a lot of money.

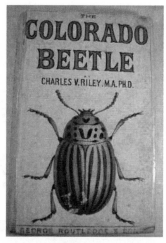

An 1877 edition of Charles V. Riley's *The Colorado Beetle, with Some Suggestions for Its Repression and Methods of Destruction*. (*The Potato Museum*)

As we have seen, taking the potato from South America and introducing it around the world created serious problems. Historically when humans move a crop, there is a honeymoon period, a time when the crop will be able to grow and flourish away from its natural enemies. Usually, the honeymoon period ends. A pest will find the plant and because of the way humans tend to introduce a crop, without much genetic diversity and hence resistance, that pest will go nuts and destroy the crop, unless pesticides keep it in check. Potato late blight and the Colorado potato beetle did nothing but exploit a resource made available to them because of the actions of humans.

Because of Europe's history as a center of exploration, it has often served as the hub through which new plants find their way to other cultures. Some of the plants that moved into Europe migrated slowly across many miles and over many years, and some moved relatively rapidly. The peach came from China, was moved into Persia, the Roman Empire, England, and then the rest of Europe. We could even say that it took a meandering path to becoming an important crop. Many other food crops, such as wheat and apples, followed essentially the same path. Some crops were introduced relatively rapidly, over the course of a few hundred years. Potatoes were discovered in South America in the 1500s and found their way throughout most of Europe over the course of a century. Peanuts had a similar journey. Over thousands of years, Europe amassed crops that originated from across the world, but if the movement of food crops into Europe was slow and gradual, the movement of species into the United States was fast and intense. Europeans had developed tastes for a wide variety of crops, many of which were not found in the United States prior to their delivery there by European colonists.

In the United States the movement of plants across borders came under regulation in the late 1800s, when state laws were enacted to prevent the export or import of trees containing injurious diseases and insects. In 1912, the Plant Quarantine Act, which "regulates the importation and movement of nursery stock and other plants and plant products within the U.S. to control the dissemination of injurious plant pests and diseases," became law. This act effectively regulates the movement of plants into and through the United States, even though it wasn't created to control plants themselves but rather to "prevent dissemination of plant pests, plant diseases or insect pests."

The importation of plants into the United States made agriculture what it is today. The difference is that today people are more aware of the possible consequences of moving plants. The Japanese beetle can be a danger to landscapes, for example, and so we limit the movement of possible host plants. Prevention is

a noble course, but these insects will eventually evade our controlling hand. Nature always seems to be one step ahead of us, no
matter how hard we try.

Trees are the grandest plants in the forests, but the next two
chapters aren't about trees in forests, they're about introduced
trees that found their way to the United States because of a
desire by colonists to eat their favorite fruits. These are managed
trees that, though unlikely to take over a forest by outcompeting
native trees, take space away from our native plants just the same,
with the help of their human cultivators.

Chapter Five

APPLES AND AGE

ONE of my grandfathers, the one with the great climbing tree, came from Massachusetts. My other grandfather lived most of his life in New York and owned a small fruit and vegetable store in Brooklyn. I can still remember this store and the Chinese restaurant next to it where he would take us whenever we visited. His store always seemed full of hustle and bustle, and I still remember the bright colors of the produce on their shelves. When he retired to Florida with my grandmother he continued to work, packing fruit at a friend's warehouse until he was ninety years old. I went with him to the warehouse once and he taught me how to place apples into containers to minimize shipping damage. The apples had labels on them indicating where they were shipped from, mostly Washington, which isn't surprising as that is the state which produces the most apples. But that's not where they came from originally, not the trees at least.

The United States has a few species of native crabapples, but the apples you and I generally think of—the crispy, sweet, and juicy fruit we love to eat—originated from an area around east-

ern Turkey and western Russia. From there, apples spread west-
ward across the globe to Greece and Rome and didn't reach the
shores of the United States until the 1600s. Apples were a popular food among colonists who missed their homelands and who
needed a fruit they could store in a cool cellar for a few months.
Orchards sprang up across North America as they developed into
a valued commodity. John Chapman, better known as Johnny
Appleseed, was one of the apple's greatest supporters and helped
spread the fruit and information about its culture and care across
Ohio and nearby regions in the early 1800s. The trees that came
from Johnny's seeds were originally collected from the waste of
cider presses and thus his trees were not, as most trees are today,
propagated through a vegetative (also known as asexual) method.
The last remaining tree from Johnny Appleseed's exploits is said
to still stand in Nova, Ohio. You can actually purchase a clone of
it grafted onto a younger rootstock, which is somewhat ironic
since to our knowledge Johnny never actually planted a clone,
and was thought to have considered this practice contrary to the
ways of nature.

Today the fruit that comes from an apple tree produced from
seed is almost worthless. When you walk into the grocery store
to buy apples you don't just buy apples, you buy particular fla-
vors and textures; you buy varieties. You know that a Granny
Smith will be green and tart. You know that a Red Delicious will
be red and sweet. You know that Honeycrisp will be honey-fla-
vored and crispy. If the apple didn't have a name, how would you
know what to expect? More so than with any other fruit, we buy
apples according to variety, and the production of apples has
been industrialized to accommodate this trend and facilitate the
growth of certain popular varieties. But most people do not
know that you cannot simply plant a seed that comes from a

Honeycrisp apple and grow a tree that produces Honeycrisp apples; you need to produce these plants asexually. If you don't then all you have is an apple tree, and people don't want to buy any old apples. Of course most people who decide to plant apple seeds aren't aware of this until their tree develops fruit, which may take some time. Seeds have their own advantages though. They're the only way to get a truly young plant, as we shall see later in this chapter.

People have dozens of ways to make themselves appear younger: creams and lotions, tummy tucks and facelifts. All to make it look as if time hasn't touched us, but in our hearts we know it's all superficial. Despite thousands and even millions of dollars spent, our bodies will never be eighteen or fourteen or twelve again. It just can't happen. It's not quite the same for our friends the trees though. There are ways to manipulate time and turn back a tree's clock, methods that just don't work with people. How else could we possess a clone of a tree planted by Johnny Appleseed? And most of these methods cost only pennies per tree—how's that for efficient? But to understand these methods, it's important that we get a feeling for how age works in trees, which is quite a bit different from how it works in humans. Age in a tree depends on many factors, one of the most important of which is how the tree was propagated. So, let's take a look at three apple trees, each propagated using a different method, and analyze how they grew, and aged.

The first tree we will observe came from a seed. The second grew from a cutting taken from the branch of another apple tree. Our final tree grew from a bud from one tree that was placed onto the stem of a younger tree. Three propagation methods; all meant to produce a new healthy apple tree. Each method has its own strengths and weaknesses, and each represents a different beginning, and a different ending.

Apples collected by researcher David Bedford showing the diversity of these fruits. The three largest apples in the upper right are Honeycrisp, Red Delicious, and Granny Smith. The rest are seedlings. Tasty, but commercially worthless. (*Emily Hoover*)

The Backyard Apple: Seeds

A small boy took an apple seed and placed it in a pot outside his house. The seed had come from a McIntosh apple that he had eaten the previous day and he wanted to grow a new apple tree to plant in his backyard. He had liked the apple and his father told him that if he planted the seed he could grow the apples himself. The seed sat in a pot in the garage over the fall and winter. The boy checked the pot every day, but after spending much time sitting and doing nothing, he was ready to give up on the seed.

In the spring the boy had had enough waiting. No tree had grown, and his parents were getting tired of seeing the empty pot in the garage. But just as the boy was lifting the pot to heave it into the trash a speck of green caught his eye. It was a small leaf emerging from the soil within the pot. He took the pot inside and placed it on a windowsill. Over the next few days he watched the plant sprout, its top stretching toward the sun that shone through the window. The boy was happy and watered and fertilized the little seedling religiously. As the summer came, the boy moved the young apple outside onto the porch. The young tree responded by growing straight and tall that first year, so much so that by the time winter rolled around again it was two feet tall and its roots were wrapped around the inside of the container like baling twine.

When the snow started to fall the boy brought the tree inside the garage so its roots wouldn't freeze. It stayed there until the following spring when the boy's father decided that the tree needed to go either in the yard or in the garbage. So the tree was planted into a bare patch of ground in the backyard. The father thoughtfully spread mulch on the ground to control weeds and fertilized and watered the tree regularly, and the tree grew. For five years the tree pushed skyward without incident. There were some years when the leaves contracted some spots, but in the fifth year, a particularly wet year, the spots grew to cover entire sections of leaves and the tree lost most of its foliage. Sure that the tree would die, the father prepared to cut it down, but in the summer the tree sent out a new, if weak, flush of leaves, and the tree survived to see another year.

After ten years the tree produced flowers. It was almost twenty feet tall now and had grown into quite a sturdy young thing. By September many of the flowers had turned into fruit and when the boy came home from school he picked the apples with his father. Both the boy and the father were surprised at the fruit that came from this young tree. It wasn't what they had expected at all. The apples were somewhat similar to McIntosh, but they were smaller and weren't nearly as sweet. The boy and his father still ate the apples right off the tree, but they preferred them cooked up in the mother's excellent pies. Inside the cells of the tree was a DNA sequence quite unlike that of the tree's McIntosh mother. Instead the tree inherited its traits from both its mother and father and its apples reflected this combination.

For us, death through aging is an insidious beast. In the end it is the effects of the clock, and not the clock itself, that usually get us. Arteries clog, tissues cease to be renewed, toxins build up from years of exposure to the sun, and chemicals, both natural and man-made, slowly poison our systems. All of these factors take a toll on health and eventually, either together or individually, lead to a person's passage from this world. But what about a tree? What happens in a tree's body that leads to death? There is no doubt that trees have an internal clock, but nobody is quite sure whether death is programmed into it or not. Trees have life spans and it is well known that certain trees will live longer than others. A white oak may live six hundred years, a persimmon will be lucky to reach 80. An apple grown from seed may reach a hundred or perhaps a hundred and fifty years before it declines. Until recently, bristlecone pines were thought to be the oldest trees on earth. These scrubby pines may grow to be almost five thousand years old, and perhaps even older. They often grow where other plants can't, taking hold in soilless areas between rocks and eking their living out of practically nothing. In the Rocky Mountains and other Western ranges these pines are rarely found below six thousand feet, growing only a fraction of an inch a year in diameter. Few pests bother them and very few other plants can tolerate the conditions they live under to compete with them for space. This allows the pine to struggle on over years and years with age its primary enemy.

Despite the incredible age to which the bristlecone pines can live, there is an older tree. In 2004 the oldest known living tree was found in Sweden; a Norway spruce that colonized the region very shortly after the glaciers receded. Though the top portion of this spruce is only a few hundred years old, the underground portion has been living for over 9,500 years. This tree has survived for so long by growing new shoots as the old ones die off, sacrificing living tissues that can't handle the size to which they've grown with young tissues that can.

The apple tree that our little boy grew doesn't have any hope of reaching the age of the bristlecone pines or the spruce in Sweden. Nonetheless, if the boy cares for the tree properly there is no reason to believe that it won't live through the boy's lifetime, and perhaps his child's lifetime as well.

Trees age in a variety of ways. Like animals, they usually go through a juvenile stage where they are attempting to gain size, followed by a mature stage where their primary goal is to reproduce. In some plants the juvenile stage is very easy to see. The maturation of English ivy is distinct. Juvenile leaves contain three to five lobes, while fruiting, or mature, portions of the plant have leaves that are more oval, even diamond shaped. Most plants aren't quite this cooperative in revealing when they shift from a juvenile to a mature stage, and so the only obvious way to tell that a plant has become mature is by looking for the presence of flowers and fruits. Apple trees grown from seed reach maturity sometime between five and fifteen years after they sprout.

After a tree reaches maturity the bulk of its available resources are earmarked for reproduction rather than growth. Size, both

Cross sections of tree trunks healing over. The trunk on the left has successfully grown around an injury while the trunk on the right has been unable to do so, exposing its wood to insects and disease. (*Chad Giblin*)

height and circumference, will continue to increase, but at a slightly slower rate than before. Meanwhile physical damage from errant lawn mowers, weed whips, and hungry deer that has affected limbs and roots will accumulate. Just as humans amass scars over the course of their lives, so do trees, but with trees the scars are often much more debilitating. So often we hear about trees healing themselves, but trees don't "heal" in the sense that we understand. Unlike an animal's body, which will heal when wounded, damage to a tree never truly goes away. Instead, scars build up and potentially affect a tree's health over time.

When a tree is damaged by a scrape or a cut it will try to close the wound by growing around it without actually healing the wound itself. Over time, usually many years, the wound will be sealed over. But when a tree seals over a wound, the tree hasn't "fixed" the damage; instead the tree has simply absorbed the damage as its girth has increased. If the damaged area never developed a disease problem, then the tree may well absorb the damage without subsequent effects, beyond the energy it needed to use to close the wound. On the other hand, if the wound had become infected by fungi or bacteria, then the wound has become a time bomb. Though the tree trunk may appear perfectly healthy on the outside, inside the tree will develop what is known as a reaction zone around the infection. This reaction zone is a collection of cells that surround the infection site and that may or may not stop the infection from progressing into the rest of the tree. Infections in trees are insidious. Sometimes they kill the afflicted tree quickly, but more often they linger, eating the tree from the inside out slowly over many years. It sounds like a long and lingering death, but it is really what both the tree and the pathogen prefer. After all, the fungi or bacteria doesn't want the tree to be killed too quickly; it needs a lasting source of food. And the tree certainly doesn't want to die quickly. It wants as many years as it can to send out seeds and reproduce. In fact, the presence of an infection often signals trees to reproduce. Trees tend to flower and fruit more profusely when they are under stress, a simple mechanism that they have developed for spreading their DNA when death seems close. Once upon a time gardeners actually recommended smacking trees with baseball bats to encourage them to flower. An interesting idea, but not a particularly good one, as this type of abuse can damage the tree's vascular tissue and lead to an earlier (if more beautiful) death. I doubt that a boy who loved a tree that he reared from a seed would ever consider using his Louisville Slugger on it.

Over time, the boy who planted the tree went off to college and, later, moved to another city for a job and a family. The father and mother eventually sold the house and moved into a more manageable space for an older couple. By this time, the tree had grown to be almost thirty feet tall. The new family that moved in enjoyed the apples the tree produced and gave them away to friends and visitors, as did the family that moved in after that, and the family that moved in after that. As the tree aged it began to grow more slowly and diseases appeared more frequently on its leaves, flowers, and fruit. Insects started to attack the fruit too. Apple maggots and codling moths made their homes inside the waiting apples, infesting them before they could be picked. The owners of the house learned to always cut the fruit open for examination before eating, to make sure there wasn't any extra protein lurking there.

Eventually the house where the tree grew was abandoned, but the tree lived on. When it was young, the boy and his father had pruned away dead and damaged limbs, as well as limbs that rubbed against each other, or grew too low and blocked the father's mowing route. Some of the subsequent owners had also pruned the tree regularly, and some hadn't, but now, with no one in the house, the tree sat alone and forgotten without any care whatsoever. The tree was over fifty feet tall now. Cracks appeared at limb junctions where too much fruit weighed the limbs down. Inside these cracks various diseases grew, using the tree's remaining healthy tissues as their food. Limbs died. New growth simply didn't emerge from them in the spring. And then, one spring, there was no new growth except for a lonely sprout that came up from the base of the trunk.

As a mature tree ages, a number of things happen to it that may play a part in its ultimate demise. Certainly the accumulation of injuries and disease doesn't help the tree's chances of survival, but there are other things happening too, things that the tree has even less control over than these wounds and diseases.

Every year, new growth forms on growth from a previous year that, in turn, was produced on even older growth. This continuous stacking of new growth on top of old is necessary for the tree because leaves flush from new growth, and leaves are needed to draw water up the tree's stem. Water is needed for most of the tree's functions, from photosynthesis to transporting nutrients like nitrogen and potassium to the places where they are needed. Though leaves are required for trees to survive (except for a very few instances where stems take on the responsibilities of the leaves), they are extremely susceptible to environmental conditions, particularly cold, and so, in temperate climates, leaves will be sacrificed once a year (or every few years, as with evergreens). These leaves are voluntarily sacrificed so that important nutrients can be collected from the leaves before they are shed. As stated by Catherine Smart, a leading researcher in the field of leaf senescence, a leaf's death is its "last will and testament . . . in which bequests of nutrients are made to the rest of the plant." If the tree simply allowed the winter to take the leaves from it, then many valuable nutrients would be lost. This sacrifice marks the entrance of deciduous trees into dormancy. When trees go dormant their physiological processes come to a virtual standstill. Yes, there is still some activity under the ground, and even above the ground, but growth has ceased, and all of the tree's parts that could be damaged have either dropped, or are waiting for warm weather. When spring comes around again the buds that were formed the previous year will start to grow, and as they elongate leaves will extend into the air. If a tree does not have the resources to extend new foliage then that tree will be unable to survive.

Leaves have, on their surface, small pores called stomata. If you cared to, you could actually trace a line from the stomata on a particular leaf through the cells of the stem and into a particular root. There is no communal pool within a tree from which leaves gather water and nutrients; rather, a stem is like a collection of very narrow straws all tied together. Leaves sit at the top of the straws and roots at the bottom. When water evaporates off the surface of a leaf, a process called transpiration occurs, and water is pulled up one of the straws from a particular set of roots. Now, this is oversimplifying things a little, and there can be some sharing of water between sections of the stem that are close to each other, but by and large, this is what happens. In fact, when one portion of a tree's root system is damaged, this damage will often manifest in only one portion of the tree's canopy, the portion whose leaves can trace their vessels to the roots that have been wounded.

As a tree ages, its leaves must draw water through a longer and longer straw. Think about the pressure you would need to suck milk through a foot-long straw. Now imagine the pressure you would need if the straw were 30 feet long. It quickly become obvious that as a tree grows it will have more and more difficulty supplying its extremities with water. This lack of water results in a few negative consequences for the tree. First, pores in the tree's leaves will need to stay closed more often. Stomata are only open when there is a favorable environment for pulling water from the ground and into the leaves. As it becomes more difficult for the tree to draw water up its stem, stomata will stay closed more often. Besides allowing water to evaporate, stomata also allow carbon dioxide to come into the plant, for photosynthesis. As a result, when the pores are closed, less photosynthesis occurs and the tree is less able to produce the carbohydrates it needs to live. Second, water is needed by the tree for leaf expansion. Cells in the tree's leaves expand through water pressure. If it is difficult for the tree to get water to its leaves then there will be insufficient pressure for the leaves to expand properly.

By getting taller every year, trees grow themselves into a tough situation. They need to grow to live, for without new growth new leaves will not be produced and the tree will be without the leaves' sugar. But at the same time, by continuing to grow, trees make it more difficult to provide themselves with the water and nutrients they need to transport water from their roots to their leaves. Though scientists have yet to identify exactly what it is that kills old trees, this catch-22 is one of the reasons that trees become weaker over time.

Some trees circumvent the conundrum by reducing the distance and distractions between their roots and their leaves. Take for example the oldest known tree in the world, that Norway spruce I mentioned previously. How could this tree possibly live to be over 9,500 years old without growing itself to death? The answer is simple. The tree's top dies off every few hundred years, and the root system sprouts a new trunk, an occurrence very similar to coppicing. By sprouting a new trunk, the tree is able to channel its resources into a system uncompromised by distance, and that is able to draw up and use the water and nutrients from the roots.

Besides threatening the tree's ability to draw up water, aging has deleterious effects on many other parts of the tree; some of these effects lead to death, and are best seen by examining our other trees, those not propagated from seed. Let's take a look at a tree grown from a cutting.

Trees and Stem Cells: Cuttings

On July 5, 1996, a sheep named Dolly was born in Edinburgh. At first glance she appeared to be a rather typical sheep. She had two ears, four legs, two eyes, the usual amount of wool and such, but deep inside her body she was hiding something important. Something brand-new to science. Only by literally reading Dolly's genes could you figure out why she was so unique, but once you read her genes, and those of her genetic mother, it suddenly became evident that Dolly was indeed quite special. Because, you see, Dolly was the first mammal to be cloned from an adult cell. This was a significant occurrence because it proved that humans could produce an animal exactly like its parent and opened the possibility of producing livestock that was all exactly the same, potentially greatly simplifying animal production.

Animals are cumbersome creatures. Each one has its own personality and is slightly different from its brothers and sisters and so requires slightly different care. For people who grow animals for milk or slaughter, producing herds of sheep, cattle, or whatever in a uniform manner is attractive because then they could predict such things as exactly what diseases the animals would be immune to and what they would be susceptible to, exactly how long and how much food it would take to fatten them up appropriately, and exactly how much they would be worth. In other words, they would become predictable, and uncomfortably similar to the trees that surround and decorate our houses today.

Trees grown from seeds are too variable for most commercial nurseries. It's almost impossible to predict how large a seed-grown tree will get, how good its fruit will be, how resistant it will be to various diseases, and how its canopy will appear. So, despite the fact that planting a seed is the easiest way to reproduce most trees, this is usually the last resort for most plant prop-

An elm cutting as it develops roots. Cuttings are the preferred way to propagate most plants because these cuttings will be the same genetically as their parent plant. (*Chad Giblin*)

agators. Seed planting is fine as far as it goes, but seeds produce trees that can be quite different from one another as our boy learned with his apple tree, and so nurseries shy away from using seeds. Instead, when you go to a nursery or garden center you will find yourself paying top dollar for a particular cultivar that has been selected to have all of the traits that you find desirable, such as disease and insect resistance and regal form and structure. These cultivars are all produced from a single parent without the use of seeds. Fortunately for the nursery, there are several ways to reproduce a tree asexually. In other words the tree will be produced without sex and the offspring will be exactly the same as the parent. Just like Dolly, but with plants it's a lot easier. Let's take a look at a tree grown from a cutting to see how that plant ages.

The cutting was taken in the spring from a young twig near the top of a tree that bore Delicious apples. It was about six inches long with five leaves. The lower two leaves were stripped off the excised stem with a flick of the propagator's wrist, and the base of the cutting was placed into a mixture of talc and a plant hormone, Indole-butyric acid. Then the cutting was placed beside other cuttings in a bed of coarse sand, its remaining leaves sitting above the surface. Every ten minutes a sprinkler would go off to keep the cuttings moist and prevent the desiccation that would kill a cutting faster than almost anything else.

After four weeks, the cutting began to grow roots. Its bark split where the new appendages grew from the vascular tissue through the outer bark tissues and into the surrounding sand. Two weeks later the propagator came back into the room where the cuttings sat and lifted the cutting from its sand bed, delicately knocked the sand off the plant's roots, and planted the new tree in a small container with potting soil. Of all of the three hundred stem cuttings that were placed in the sand, only this one and three others produced roots.

Unlike many other shrubs and trees such as roses and magnolias, apple trees are not usually produced from stem cuttings. It is not that apples cannot be propagated by cuttings at all; rather, the percentage of cuttings that will grow successfully is so small that it does not make sense to grow them this way. Every once in a long while you will find an apple tree for sale at a nursery grown from a cutting and not from one of the preferred methods used for propagating apples, budding and grafting, but these are few and far between.

Stem cuttings have been used for thousands of years as a method to propagate plants and are very easy for anyone to try if they're so inclined. To see stem cuttings in action the best, and easiest, way is to take a few willow branches and place them into a glass of water. Within a few weeks, or sometimes just a few days, the shoots will have developed roots. At this point the shoots can be taken out of the glass and planted. The roots will function just like the roots of any plant and viola! You will have a small clone of the willow you took the cutting from. This tree will grow exactly the same as the tree which it originally came from did and so you can predict what it will be like as it gets older. Of course, not all trees produce roots on stem cuttings as effortlessly as willows do, usually taking considerably more cajoling, but when a tree can be reproduced by cuttings then that is what a nursery worker will usually do.

Besides the ability to create exact clones of the plant you want to reproduce, stem cuttings offer other advantages as well. They are easy to care for, perhaps even easier than seeds since they never have the tiny tender shoots that seeds produce when their hard seed coat first opens. Stem cuttings can be taken almost any time of year, depending on the plant to be propagated, and some level of success can still be had, but really, besides the ability to clone the plant that you want to reproduce, the major advantage of working with stem cuttings, and any other means of asexual propagation, is that you will be able to maintain the maturity of the plant that you are reproducing. What this means is that the plant which grows from the cutting will be significantly older than a seed that was planted at the same time. Exactly how much older is dependent on a great many things, including the species of plant that you're propagating and where on the parent plant the cutting comes from. Regardless of this, the newly produced plant will flower earlier in its life than a plant produced sexually would, making it possible to sell it sooner than plants produced by seed (we all like flowers on our plants!). It will also, unfortunately, have had its life clock begun earlier and so this plant may live a shorter life than its seed-grown brethren.

WHEN Dolly was born she came from a cell that originated in the mammary glands of her mother. This cell served as a stem cell: a cell that could be made to produce any type of new cell that was desired. Most animal cells aren't like this. Animal cells are generally programmed to produce only certain things. If the cells in the bodies of animals could produce new cells of whatever type was needed or desired at a particular time then we would have no trouble growing back fingers, toes, or other body parts if they were somehow cut off or otherwise lost. Obviously this doesn't happen, at least not in most animals, unless of course, you include Dolly and clones like her who came from a single cell specially treated so that it will produce every part of an animal's body. In this sense certain cells in a plant's body are like stem cells. These cells are part of the plant's meristem.

Meristems are the portion of the plant where growth takes place. The tip of the shoot, the tip of an expanding leaf, and the interior of the bark where new vascular tissue is laid down and which widens the trunk of a tree, are all examples of meristems. When a cutting is taken the propagator is counting on something innate to meristems in a plant to produce roots, and that something is totipotency. Totipotency is the ability of a cell to produce all of the other types of cells in a plant. Stem cells are, by definition, totipotent. Within plants particular types of meristems may simply appear when they are triggered by an environmental queue. For example, a willow placed in a glass of water will produce root meristems from cells in its stem with little encouragement. Most plants, though, need to be somehow stimulated to produce this extra growth. This stimulation comes in the form of plant hormones, applied by humans, that prompt the plant to produce new growth where none grew before.

Plant hormones have a long and involved history with humans. There are a number of plant hormones that affect plants in a variety of ways, but the ones most used (perhaps overused) by humans are the plant hormones known as auxins. These hormones are more than a little bit analogous to the hormone testosterone in humans in that they promote growth. And like testosterone in humans, this encouragement of growth in plants is not always to their benefit. First discovered and researched by Kenneth Thimann and others in the 1930s, a number of chemicals function as auxins. The two that are commonly used for encouraging cuttings to produce roots are Indole-3-butyric acid and Naphthyl acetic acid, neither of which are produced in any quantity in nature, but when they are applied to a plant's stem they behave just as a natural hormone would and inspire growth. Though artificial auxins can be used for good, just like steroids, they can also be used for ill. The most used herbicide in home lawns, 2,4-D (dichlorophenoxyacetic acid), is an auxin closely related to those used for encouraging root growth. It was a component of the now infamous Agent Orange used during the Vietnam war along with another auxin, the now illegal 2,4,5-T (trichlorophenoxyacetic acid). Despite their ability to kill plants, at low doses and applied properly to a cutting, these herbicides can also be used to encourage root growth in stem cuttings.

The small tree grew into the pot, quickly filling it with roots and sending up three new shoots from the buds that sat on the stem just above the leaves remaining on the cutting. All but one of the new shoots was cut off. Now the nutrients taken up by the roots were funneled into the single remaining shoot which responded by growing fast and straight. By fall when it lost its leaves the young tree was almost two feet tall. The man covered it and the other plants growing in containers with a polyethylene tarp followed by a layer of hay and then another polyethylene tarp. The new plant spent the winter chilly, but protected from extreme cold by its covers. In the spring, as the new plant was warming up and starting to demand water for its expanding branches the man took the coverings off and irrigated his stock. The plant was placed with other trees and arranged on a nursery bed where it would live for the next year doing nothing but growing. Insecticides and fungicides were applied to the apple trees regularly to ensure that pests didn't compromise the growth of the young plant. New branches and leaves emerged along with a few flowers. The man and some of his helpers removed the flowers. If they were allowed to be pollinated and grow into fruit it was possible that the weight of the apples would bend and perhaps even break the young tree's trunk.

One year after the first flower appeared on the apple tree it was sold to a retail garden center. This year the flowers were not removed and so the beautifully flowering tree was quickly purchased by a young couple who wanted to plant an apple in their backyard for the beauty of the blossoms and the fruit in the fall. They planted the tree quickly and, though the tree was purchased with blooms, the flowers were blown off by a windstorm and the tree produced no fruit the first year it was planted. The following year the tree flowered and fruited in abundance and the couple was very happy with their purchase.

THOUGH the early maturity of a cutting propagated plant makes it attractive to people who want flowers, there is an inherent problem with plants that are propagated through vegetative means alone, something called, ominously enough, clonal decline. Consider the tip of the shoot of a plant. This shoot will include a meristem from which new growth will emerge in the form of a stem that can be used as cutting. Once this cutting becomes a new plant, the tip of this new plant can be cut off yet again to form another new plant, and so on. In fact, there is no limit to how many times a young stem can be removed which includes this very same meristem which will then produce a new shoot for a cutting. In other words, though a plant grown from a cutting may seem young, the meristem of this little bit of plant may actually have been around for quite a long time. In fact, this little piece of plant material may easily be forty or fifty years old, though the plant in your garden has existed for only a year or two. Over time that little piece of plant will accumulate muta- tions, or problems with its DNA, which may or may not affect how the plant grows.

Think of all the different shoots growing from a stem of a plant. At the tip of each shoot is a meristem from which new growth will emerge. As the stem grows, however, the meristem will amass mutations. In other words, as the stem grows chances are that it will not be genetically exactly the same as the other stems in the plant. The mutations in the meristem can be caused by a number of things. Sunlight, a buildup of toxins, even phys- ical contact with a nearby building could trigger the DNA in meristem cells to change. This difference can manifest in a vari- ety of ways, most of which aren't obvious. In fact, most are so slight that neither you nor the tree will ever see any difference; however, sometimes changes are surprisingly obvious. These problems may accumulate over time and so, though the plant that you have is apparently young, the meristem from which it was propagated has accumulated many genetic problems and the plant will perish long before you would expect it to. Alternatively, the mutations could end up being a good thing, at least from a human point of view.

There is a type of apple you can purchase called Fireside. It's a good apple. First introduced from the University of Minnesota in 1943, Fireside has a big, sweet, firm fruit, but it isn't the most attractive color in the world. It's red and yellow streaked so you're not quite exactly sure when it's ripe, and it's certainly not the bright red, lustrous apple that consumers like to buy. In 1956 Tom Connell was harvesting the trees in his family orchard when he discovered that the apples from some of his Fireside trees were bright red. In fact, a number of his trees had red fruit. Other than the redness of the apples there was essentially no difference between the fruits from Fireside and those that Tom Connell had planted. However, that difference was enough to make these fruits significantly more attractive than Fireside apples and so Tom named the variety that came from his trees Connell Red and propagated it.

As with many of the varieties of apples that adorn supermarket produce sections, Connell Red came from something known as a branch sport. The difference between Connell Red and Fireside was the result of a mutation to one of the meristems on a branch of a Fireside apple tree that was removed, propagated, and eventually grown into the tree that Tom planted. These mutations are common in many types of plants and, if they produce something aesthetically attractive, perhaps variegated leaves or a different color flower, or, as with in the case of Connell Red, a different colored fruit, then they will be propagated and a new variety will be produced. But this variety will only be able to be produced vegetatively. Seed from these plants won't retain the mutated characteristic which makes the plant valuable. Besides Connell Red, many other apples come from branch sports too. When you think of a Delicious apple you tend to think of a bright red apple with a sweet interior, but that isn't what the original Delicious apple looked like. In fact, the original Delicious apple had green stripes on it and wasn't as shiny as the Delicious apples that you buy in the store today. That's because what you buy in the store is a branch sport of Delicious that was propagated because of its superior color. But the mutations that occur to meristems which form branch sports don't have to be attractive. In fact, they can be deadly.

Strawberries, like apples, are grown asexually. Though you almost never see strawberry varieties listed in the grocery aisle, growers are well aware of the various types and select them based on traits such as berry size, flavor, resistance to disease, and harvest time. If you plant Chandler strawberry, for example, you can expect medium sized berries with excellent flavor, but only moderate yields. Ventana, on the other hand, produces excellent yields but has only good flavor. Strawberries are reproduced by a sort of division, where runners, which are a type of ground-hugging stem, are allowed to produce roots and are then divided off of the main plant and grown into new plants. In other words, strawberries are produced through a process roughly similar to stem cuttings. When a strawberry cultivar is reproduced by runners, all of the daughter plants will have exactly the same characteristics as the mother plant, including the mother plant's age. If there is a problematic mutation in the mother plant from which the new cultivar originally came, this will be carried on to the offspring.

There is, in strawberries, a disease affecting some varieties and
not others called June yellows. This
is a disease whose cause is not clear.
It does not seem to be related to a
simple virus, bacteria, or fungi.
Instead, it seems to be a genetic disease triggered by age. Every plant
of a variety which has June yellows
is a time bomb. And since every
plant is propagated asexually, and
are almost exactly the same age, all
of the bombs are set to go off at
once. Typically June yellows will
not be seen on a strawberry variety

A strawberry with June yellows.
This disease will lie dormant in all
the plants of a particular cultivar
for years and then manifest all at
once. (*Jim Luby*)

at all and then will rapidly appear
over the course of a few years on most of the plants of this variety, regardless of where in the world they are. Breeders of strawberries never know whether the variety that they have created
has June yellows until years after the variety is released, and then,
Wham! There goes that variety of strawberry. One of the most
famous strawberry varieties to have had this occur was
Blomidon which, in the mid-1980s, was considered to be one of
the up-and-coming cultivars sure to dominate the market in the
Northeast and Midwest. It was performing well across much of
the United States when June yellows started to appear in the late
1980s and early 1990s. By 1996 this variety was little more than
a memory, disappearing as age caught up with it.

Though you would expect a cutting from a tree to retain the
age of the tree it was taken from, there are ways to take cuttings
that are younger. You see, unlike animal's bodies, there are actually older and younger parts of a tree's body. This seems a little

counterintuitive at first. After all, isn't every part of a tree the same, and didn't I just finish talking about a disease of strawberries whose age retention resulted in a cataclysmic loss of plants? Yes, but trees are different. Strawberry stems run along the ground and don't grow vertically. And vertical growth is how a tree maintains different ages on the same plant. The youngest part of the tree is the part near the earth and the oldest part of the tree is the part at the top. OK, I know that seems backward; isn't the top of the tree where new growth is popping out and so shouldn't that be the youngest? In a word, no. Think about the tree and what it looked like when it was young, just a seedling. It would have been close to the earth. Over time it would have grown taller, but that isn't because the part of the tree that was near the earth grew upward, it was because new growth grew on top of the growth that was already there. Here's the rub, as the tree grows up it matures. The growth that has already been deposited, the part of the tree trunk near the earth, will stay relatively young while the growth near the top, though it may be newer chronologically, will be physiologically older because the tree is programmed to mature as it grows, and the only way for this to happen is for the new growth (chronologically) to be more mature than older growth. This little arrangement can have a profound impact on the age of the tree that you propagate from stem cuttings. Cuttings taken from locations closer to the top of the tree will produce trees that are physiologically older while cuttings taken from closer to the base of the plant will produce trees that are physiologically younger. Suckers, those shoots that develop around the base of a tree, are the source of the youngest cuttings you can take from a tree. As the people who planted the apple grown from a cutting find out, the early maturity of the tree they grew is offset somewhat by its shorter life.

✧

As the couple grew older they had children and their apple tree provided them with an almost inexhaustible source of apples. Their children enjoyed the apples year after year and so did all of the neighbors who frequently received apple gifts. As they grew older the kids left home for college and later jobs, and the house was empty except for the couple again. They continued to enjoy the tree, but more and more often they needed to bring the neighbors in to help them take care of the extra produce from the forty-footer.

After fifty years the tree was starting to wane. Growth was slow and the canopy wasn't nearly as dense as it used to be. The tree still produced apples, but they tended to be buggy and no one wanted to be bothered with applying chemicals. Besides, the couple considered themselves too old to continue tending the tree. Eventually they decided that they had had enough and had the tree cut down, replacing it with a hawthorn. A sad end for an old friend.

DOLLY was euthanized on Valentine's day, 2003. Hers was a short life, even for a sheep. The cause of her death is listed as a progressive lung disease and crippling arthritis, both of which are diseases more commonly found in older sheep. Though Dolly's predominantly indoor life may also have been a significant factor, some scientists have speculated that the cell taken from Dolly's mother and cloned to produce Dolly retained the age of its parent. Because the cell retained its age, they reason, her body may have been significantly older than her chronological age would indicate. Other scientists think that this conclusion is without foundation. Ultimately this question may never be satisfactorily answered, at least not for Dolly, but for trees, maybe someday we'll understand age well enough to explain exactly what goes on in their cells as they grow older.

The Modern Apple: Budding and Grafting

In almost every discipline there are superheroes. Einstein, Linnaeus, Nobel, Watson and Crick. Thomas Knight is not likely to become a household name in the near future, but he was a horticulturist and researcher who helped to develop the procedures and practices that we use today to grow our horticultural crops. Before we start talking about Knight's work with budding and grafting though, it's worth talking about his other contributions. For example, Knight was the researcher who first showed that gravity is the reason that roots don't grow up. He accomplished this by creating a revolving wheel in which to grow plants. By continuously spinning the wheel and noting which way shoots grew (inward, toward the wheel's center) and the way the roots grew (outward, away from the wheel's center) he showed that gravity is what makes a plant grow in the direction it does. He was also the first to recognize that leaves transpire, losing water through pores. By placing a sheet of glass on both sides of a leaf he demonstrated that one of the plates would become wet (the sheet on the bottom of the leaf where pores are present) and one would remain dry.

Though the practice of grafting one plant onto another had been known for a long time prior, it was Thomas Knight who first noted how grafting affects the maturity of apples and other

trees. He reported on his findings
in 1795 when he wrote to his
friend Sir Joseph Banks about the
trials and tribulations of grafting
apples onto rootstocks, noting,
most eloquently, that placing buds
from trees that were mature onto
young trees resulted in the apple
bearing fruit more quickly, perhaps
a year or two after the graft was
made. He also noted that taking
buds from very young trees and
grafting them onto more mature
trees didn't result in more rapid
fruit bearing on the stems which

Thomas Knight (1759–1838), one
of the forefathers of modern horticulture. (*Royal Horticultural Society*)

came from the young bud. To take it a step further, he took cuttings from near the base of a pear tree which, you will remember, is the youngest portion of a tree physiologically, and cuttings
from the ends of some shoots at the top of the same tree. He
then grafted these cuttings onto a different pear tree. In doing so
he found that the cuttings from the base of the first tree
remained juvenile for many years (juvenile pear trees are easy to
recognize because they are thorny) while the cuttings from the
ends of the shoots in the first tree were juvenile in the first year
and then matured and bore fruit the second year. Knight's observations provide the backdrop for our third, budded, tree.

The budded apple was handled carefully by the propagation
technician. The base of the apple was the famous M9 rootstock.

The top was Snowsweet, a new variety much anticipated by the apple industry. This was a state-of-the-art amalgam, created by modern science to produce the best tree possible. No one would mistake this apple tree for those that Johnny Appleseed planted in the early nineteenth century. This plant was designed as much as it was propagated, with a rootstock that insured high yields and a top comprised of a scion that would produce apples sure to be a bestseller.

LIKE the tree above, nearly all apples grown in orchards today are produced by budding or grafting. Seeds produce offspring that are too variable, and cuttings simply don't work well enough to be of much use. And so the modern apple grower uses a technique to place the top portion of a desirable apple onto the roots of another apple. It is easy for a breeder of apples to select an appropriate scion, or top portion of the tree. Trees producing tasty apples that are attractive to look at make the best selections for a scion, though other characteristics such as disease and insect resistance are also important. Selecting what apples to use as rootstocks is more difficult, and once upon a time seedling trees were used without knowing their good or bad traits, but today we recognize that certain rootstocks have more value than others for a variety of reasons.

Snowsweet is a selection of apple that was introduced into the trade in 2006. This apple variety comes from a single seed which was part of a Minnesota breeding program. Planted in 1970, the seed came from a cross between Sharon, a sweet, firm apple that may taste somewhat woody, and Connell Red, a large, sweet apple that you may remember was a branch sport of Fireside.

Dwarf apples on M9 rootstocks. These apples will never grow higher than about 15 feet tall because they have been grafted onto a rootstock what will keep them short and manageable. Across the United States and the world very few apples are grown on their own roots. (*Jim Luby*)

Then the tree was watched and its apples tasted along with the apples from hundreds of other trees planted at the same time the original Snowsweet was. Year by year trees that produced inferior apples were culled until, in 2006, thirty-six years after being planted, researchers released this apple to growers. The selection was based on one thing, the apple. With a fantastically sweet flavor, and a beautiful white flesh that wouldn't quickly become brown when exposed to air like your average Delicious or McIntosh, this apple was bred to succeed in the modern marketplace.

Apple rootstocks, the bottom of the apple trees planted in most orchards, have an interesting history very unlike that of the apple scions which they support. One of the most commonly used rootstocks across the world is M9. M9 does not produce a tasty fruit itself, but does have some properties that make it a favorite root system of growers. This apple originally came from a chance seedling in France and was selected as a dwarfing rootstock in 1828. It was originally known as "Jaune de Metz" before Sir Ronald George Hatton collected it in 1912, along with a series of other rootstocks, and gave it the somewhat less than inspired, but easily recognized name that it now holds. Hatton is famous for collecting and subsequently standardizing the names of apple rootstocks; unfortunately he was not particularly creative in selecting names and so he simply used M to signify the East Malling Research Station in Kent, England, and the number 9 to signify nothing more earth-shattering than the location of the tree in Hatton's test plot. Most of the rootstocks used today are from the M series though there are a number of others originating from other programs that have emerged since Hatton collected his rootstocks including Ottowa, Budagovsky, and Kentville. Rootstock selections from these programs have likewise uninspired names such as O.1, Bud.9, and KSC.3.

The rootstocks onto which scions are grafted must all be the same genetically. Because of this apple rootstocks need to be propagated vegetatively. Seedlings, because they have two parents, are too genetically diverse and won't produce the uniformity needed by the orchardist. We know that stem cuttings won't work well, so what's left? The most common method used today for the production of rootstocks is layering. This method takes a great deal of effort, but it is also very effective. Layering may be done in many ways, but all of these include mounding soil around a stem or a branch that is bent to the ground. Roots will then develop on the stems where they are buried under the soil and, after a good root system is developed, the stem with its new roots can be removed from the ground and a scion can be grafted onto it.

The rootstocks that Hatton collected, and the rootstocks developed by others, are important to apple production for a variety of reasons, but the greatest one is that these rootstocks offer the scions that are budded or grafted onto them a distinct advantage over other trees: small size. Large trees are not desired by most apple growers for a few very important reasons. The first is that large apple trees also tend to be wide apple trees. Wide apple trees cannot be grown at a high density because they would grow too close to each other and shade each other's limbs excessively, reducing production. Small apple trees, in contrast, can be planted more closely and, through this higher density planting, be more productive. But dwarfed trees don't just offer small size. Dwarfing rootstocks also encourage earlier than normal flowering and fruiting, often resulting in the tree producing fruit a year earlier than a nondwarfing rootstock. Furthermore, small trees mean workers can harvest without using ladders, an advantage that translates into a not insignificant reduction in liability insurance for growers, not to mention speed in picking. Hatton offered apple production a way to reliably predict what a particular rootstock would do.

Though Hatton standardized existing dwarfing rootstocks, he did not introduce dwarf apple trees to the world. Dwarf apples have been known for over two thousand years, and the use of dwarfing rootstocks has been known for over five hundred. The first report of a dwarf apple comes from Greece where Alexander the Great sent back a dwarf apple to Theophrastus from his visit (otherwise known as conquest) to Asia Minor. Theophrastus was then directing the Lyceum, a center of learning in Greece. He wrote about this tree and noted that it had probably been grown for many years in Asia. In the fifteenth century, budding and grafting techniques had developed to the point where gardeners would graft the shoots of desirable apples onto dwarfing rootstock so that they could create small, elegant trees.

Besides small size and precocity, dwarfing rootstocks may also offer resistance to various diseases and insects that attack apples such as the wooly apple aphid and fire blight. A series of root-stocks called the MM series (for Malling Merton) was bred specifically to be resistant to the wooly apple aphid, an insect that attacks apple trees, particularly in Australia and New Zealand. M9 isn't special in terms of resistance to insects and diseases, it just makes for a small precocious tree.

It's a cutthroat world out there, and only the best apple culti-vars ever make it to the marketplace. The newer cultivars that orchardists are planting today are superior in almost every way to the apples that were planted one hundred, sixty, or even twenty years ago. Our apples today tend to taste sweeter, last longer in storage, and be more aesthetically pleasing than older varieties. In fact, the standbys that so many of us are familiar with today, Delicious, Granny Smith, and McIntosh, are already falling from favor and are likely to be gone to the point of obscurity in less than a hundred years. Few have heard of Rambo except as an action movie character, but Rambo is the name of the apple that Johnny Appleseed grew and distributed across Ohio (though one could easily argue that the apples that grew from the seed that he planted were not really Rambo because they were grown from seeds). The apples coming into favor today include Honeycrisp, Jonagold, and Empire. And even though these apples are popular today, apples like Snowsweet and Ginger Gold may take over from them in a few years.

It's difficult for an apple producer to change out their entire orchard every time a new cultivar comes along. What do you do if you've got a field full of Delicious and everyone wants Honeycrisp? Who wants to tear a thousand trees out of an acre of land? It's a tough situation, but fortunately for apple growers they have a much less costly alternative. Instead of removing all their trees and replanting new trees with new rootstocks, they can do something called topworking. Topworking is a labor-intensive process, but it beats pulling whole trees out of the ground. When an orchard is to be topworked, all the branches from all the trees in the field are cut back to only four or five stubs. If you didn't know better you'd think that the whole field was being prepared for removal, but it's not. Onto each of the stubs created by this heavy pruning two twigs of the new cultivar to be grown are grafted. This grafting brings a brand-new cultivar to the field without the removal of the old trees. Of course it will take a year or two for the field to produce, but by avoiding replanting a whole field, costs will be much lower than they would without topworking.

Our grafted tree grew well, and in its third year it produced a crop. It was tied to a stake, as most trees grafted onto M9 rootstocks are. These rootstocks root weakly and don't mine deep into the earth like seedlings or nondwarf rootstocks do. If the tree hadn't been connected to the stake it surely would have fallen over under the weight of the apples, but with its branches tied up it remained healthy. Over the course of the year it had been sprayed a dozen times. Fungicides and bactericides were applied to control apple scab and fire blight, and, later in the season, insecticides to kill codling moths and apple maggots trying to tunnel into the apples. It had also been fertilized every year to encourage a better crop the subsequent year. The tree pulled nitrogen, phosphorus, potassium, and other elements from the soil to grow its apples and, when these apples were harvested, those nutrients left the field and needed to be replaced.

In the next field over trees were literally falling apart. Honeycrisp apples, budded onto B9 rootstocks over fifteen years ago, were starting to break off at the graft union. Too low in the canopy for the tree to be topworked, every tree that was broken would need to be removed. Because of this problem the entire field might need to be replaced.

As our grafted tree discovered, one of the biggest problems in any fruit orchard is that of delayed incompatibility. Some rootstocks simply do not work with some varieties. Most apples can be grafted reasonably well with most rootstocks, but some, for whatever reason, simply do not hold together over time and the only way to find out whether a particular rootstock and variety don't work together is to graft them and wait for something to happen. Unfortunately, by the time you figure out that there's a delayed incompatibility problem, many orchards have already been planted. Delayed incompatibility may show up two years after grafting, or twenty. And it's not only fruit trees. Many trees in the landscape are grafted and, in some cases, delayed incompatibility occurs. Often it is these trees that fall over in a heavy storm. Though we can't necessarily see the problem clearly from the outside, these trees can be significant hazards because of the propensity of the graft to fail during a windstorm and fall on something that you don't want it to, like a car, house, or even a person.

When an apple variety is grafted onto a rootstock there is a pronounced effect on the physiology of the tree. For example,

A row of trees with weak graft unions. These trees are over thirty years old and show signs of delayed incompatibility between the rootstock and scion. (*Jeff Gillman*)

water has a more difficult time making its way from the roots to the leaves of the plant. Additionally, if the graft is planted too low roots will develop and extend from the scion into the ground, making the rootstock lose some of its dwarfing effect. Alternately, a second problem may occur. Suckers may develop from the rootstock, and they are definitely not beneficial because they will not grow the variety that is desired; rather, they will grow apples which come from the rootstock from which they sprouted and are generally small, sour, and nasty. Coppicing is not a reasonable management technique for apple budded or grafted onto a rootstock.

Our tree produced fruit year after year. In the spring, bees would be brought into the orchard to pollinate the apples, ensuring that the flowers would turn into fruit. Scheduled around visits from the bees, chemicals were applied to keep it free of pests. The tree provided a steady flow of pristine apples that consumers would pay a premium for. The quantity of apples that the tree produced increased only slightly every year. In the spring, after the flowers had been fertilized and while the fruits were growing, laborers would come into the orchard and pick off the unripe apples to encourage the remaining apples to grow big and juicy and to prevent any single limb's apple load from becoming too heavy and shearing it off. Though many branches were produced, the tree never grew taller than the migrant laborers picking the fruit could reach. If it weren't for the stake holding the tree up, the branches might have fallen even lower.

After twenty years of producing fruit for the farmer the tree suddenly lost its value. Not because of a decline in the health of the tree. It was still growing and producing fruit, and not because of pests. They had been and would continue to be a problem, but the pesticides took care of that. Rather, the tree that was once a state-of-the-art construct was now passé. No one wanted Snowsweet any longer. It just wasn't in demand. So the farmer cut off the limbs of the Snowsweet near their base, and inserted twigs of a hot new cultivar that was sure to do well in the marketplace.

APPLES are one of the horticultural crops on which we apply the most pesticides. There are a number of afflictions including apple scab, fire blight, codling moth, and apple maggot that can infest the apples and make them essentially worthless. Certain types of apples are more resistant than others to these pests and the diseases they carry, but none is immune. Organic farmers try various methods to keep pests from trees, but they often end up using chemicals to control the pests as well, albeit organic chemicals such as copper sulfate and pyrethrum which can be quite toxic to people and the environment in their own right. Without human intervention apple crops are small and more appropriate for the cider press than for eating and apple pies. Only by constant attention and, before that, careful breeding and handling could our tree have achieved the productive life that it led.

Chapter Six

THE SHORT LIVES
OF PEACH TREES

My family moved to Pughtown, Pennsylvania, to experience "country life," and one of the first things that my father did was to plant peach trees throughout the yard. In the end he planted over two hundred trees spread out over about an acre of land. Four years after planting we had peaches practically coming out of our ears. We ate peaches, we ate peach pies, we sold peaches by the roadside and to the local grocery store. Even the horses and chickens ate peaches. All this and there were still plenty of fallen peaches on the orchard floor for the Japanese beetles and yellow jackets to munch on.

Most of the trees my father planted were set out in 1978 and 1979. By the time I left the house for college in the late 1980s many of the peach trees were starting to wane. Today, thirty years later, only two of those two hundred trees are still alive. But as quickly as the peaches that my father planted faded, I learned that they didn't go half as quickly as peaches grown in the Peach State of Georgia do. The story of why these trees die so quickly is a simple one: Of all the fruits introduced into the United States over the last five hundred years or so, the peach has been pressed into service the most rudely, encountering some of the most aggressive pests America has to offer. But they aren't the only fruits to suffer at the hands of American insects.

When the first explorers came to the New World they were amazed at the amount and diversity of plant life. Early in the history of the Americas, explorers were sent to catalogue these plants and send them back to their European sponsors. John Bartram, who became the King's Botanist in 1765 and whom Linnaeus called "the greatest natural botanist in the world," was perhaps the greatest plant explorer on American soil. He set a goal early in his career to document the flora of the New World and, while he ultimately failed, he made a strong start. Bartram was a close friend of Benjamin Franklin, and together they founded the American Philosophical Society, which was an outlet for early scientific work in the New World. Unlike Franklin, a remarkable jack-of-all-trades, Bartram was concerned mostly with plants and, to a lesser extent, insects and other wildlife.

The collections Bartram made, called Bartram's boxes, were sent to England for distribution, but these boxes represented only a small portion of the plant materials reaching that country and other European destinations at the time. Explorers in Asia, South America, Australia, and Africa sent back new plants, many of which ended up being planted well outside of their usual environments. The result was that some plants fared quite well and others performed poorly. The American elm, which we will investigate in a following chapter, performed poorly in Britain because of insect and disease pressures, as well as the fact that the environment wasn't right for this tree to reach its potential. Other trees, such as the black locust, prospered. Because of the sordid history of the United Kingdom and the many cultures that have visited over the years, it is difficult to tell which trees originally came from that isle and which originated elsewhere. Species from all over the world have adopted Britain as a home and for some plants, such as pear, it is unclear whether these species originated there or were introduced from somewhere else very early in, or perhaps even before, recorded history.

England was only one possible destination for plants coming from the New World. All across Europe countries were hungry for new and different plants for food and for decoration. From its discovery until today America has provided many species that now decorate foreign shores, sometimes to the detriment of European crops. One of the great staples of European tables, particularly in France, is wine. Though it is known as a luxury today, wine was once known as a necessity because it provided a source of liquid free from hazardous bacteria which were regularly found in urban water supplies. For this reason, as well as because of its flavor, wine was the beverage of choice across most of Europe. As good farmers, French grape growers were interested in improving their crop and so chose to test American species to compare with their native grapes. When American plants were shipped across the ocean in the mid-1800s, they were not sterilized (sterilization was not normal practice at that time) and so insects and diseases could hitch a ride across the Atlantic and infest European plants. This is the course taken by the greatest scourge of grapes ever known in Europe, grape phylloxera.

Grape phylloxera is an insect similar to an aphid, with a small, soft body. They feed on various parts of the grape plant, but are most destructive when they feed on the plant's stem and roots. For American grapes, an infestation of phylloxera is damaging but not fatal. These grapes have evolved with grape phylloxera for many years and have acquired adaptations that help them survive phylloxera feeding. The phylloxera enjoys this arrangement because it wants to feed on the grapes, not kill them. Without having been exposed to this insect before, European grapes were extremely susceptible to phylloxera and so when this insect was introduced, growers across the continent had huge losses and vines of some of the finest grape varieties were decimated. The situation was particularly unfortunate because the American grapes didn't produce particularly tasty wines, and for a while it seemed that this beverage might be a thing of the past. It is estimated that during the Great Wine Blight (the name given to the period in time in European history when French wine production was at its lowest because of grape phylloxera) over half of European grape vines were killed. From the time that the phylloxera was introduced until it was finally controlled about fifteen years later the production of wine in France dropped by half but, more tellingly, importation of wine into France was fifty times what it had been prior to the crisis.

John Bartram (1699–1777), King's Botanist and one of the great plant explorers of the United States. He was responsible for introducing many new plants to Europe. (*Independence National Historical Park*)

As the phylloxera spread, a variety of different techniques were used to control it, starting, naturally, with pesticides. Unfortunately for grape growers, few pesticides could make a dent in phylloxera's armor. One chemical that could was carbon bisulfide, a particularly nasty pesticide that poisoned not only phylloxera, but also people, and sometimes even the grapes. It was also flammable to the point that men had to follow along behind the pesticide applicator to put out fires. It is safe to say that carbon bisulfide was little better than a dangerous band-aid on a growing problem.

What ultimately rescued Europe from the Great Wine Blight was grafting. Grape phylloxera feed on grape roots, but the real damage comes not from the feeding itself but from the hormones injected by the insect into the grape vine. The phylloxera, it seems, inject hormones to make the grape vine grow abnormally. The abnormal growth surrounds the insect as it feeds and provides protection and a place for eggs to be laid. In most American vines, growth isn't altered sufficiently to kill the vine; however, in European vines, the abnormal growth is so pronounced that it actually inhibits the flow of water and nutrients in the plant's vascular system, preventing water from reaching the leaves. When European grape varieties were grafted onto the roots of phylloxera-resistant American varieties, the European grapes could survive.

One of the great debates of the time was whether the American rootstock would alter the flavor of European grapes, American species being considered inferior to European (and those of you who have tried muscadine wine may well agree). Fortunately for wine connoisseurs everywhere, it was ultimately established that using American rootstocks has little effect on the flavor of the wine.

Gardeners in the United States have been introducing plants from other parts of the world since the 1600s, but these introductions are different than those that were happening in Europe. Even while John Bartram was making his collections to send to England, other plants were being sent here for use by the colonists. Mostly food-related, these plants usually found their way to America because of a European who could not leave a cherished delicacy behind.

One of the first nonnative plants to reach the shores of the
United States was the peach. A native of China, where it had
been grown for over 4,000 years, the peach found its way to
Europe through Persia from which it acquired its species name,
persica, because of the mistaken idea that peaches originated in
that country. No one knows exactly when the peach was
brought to the United States, but sources indicate that the peach
was introduced by the French along the Gulf Coast or planted
by the Spanish in Florida sometime in the sixteenth century. The
most noteworthy plantings of peach, however, seem to have
originated from seeds brought from England and were planted
about the time the pilgrims landed. This is particularly interesting when you consider how popular the peach was in England
despite the supposed fact that King John died of dysentery
brought on by eating green peaches and ale. By the eighteenth
century there were copious numbers of peaches growing along
the East Coast of the United States. Thomas Jefferson, in particular, was a big fan of peaches, not only for their fruit, but also
because peach trees were useful as a source of firewood, due to
their propensity for losing limbs. Jefferson went to the trouble of
calculating that if you were to plant peach trees twenty-one feet
apart on five acres of land you would have enough dead wood
for a fireplace in a year, and this was at a cost of only seventy
peach pits (needed to replace the dead trees).

When peaches were first planted in the South, very few varieties would fare well. Spanish varieties were planted in Florida
from the days of the earliest settlers, but peaches weren't shipped
from the South to northern markets until the 1880s. The natural range of peaches is above Florida's latitude; they are native to
a more northern climate, something similar to what we find in
New Jersey. In fact, in 1890 the greatest peach-growing district

in the country sat in the tri-state area of Maryland, Delaware, and New Jersey.

Peach trees are like many other trees in that they have what is known as a chilling requirement. This requirement is what makes certain trees appropriate for certain areas and not others. It is, for example, why a Jersey Queen peach will produce peaches in New Jersey but not as many in Georgia, and why a June Gold peach grows well in Georgia but not New Jersey. A chilling requirement means that peaches need to experience cold for a certain amount of time every year in order for them to flower and produce fruit properly. The cold works a little like one of those pop-up buttons on a turkey that pops exactly when the turkey is cooked enough to come out of the oven. Trees with a chilling requirement won't bloom until they experience a length of time below a certain temperature (about 45 degrees Fahrenheit), but above the freezing point. The tree has this mechanism in place so that its blooms can avoid cold weather. Without it a warm spell in January or February might entice its blooms to open only to be decimated by a winter that had not yet ended. In the United States we have a wide range of climates and a wide range of chilling that can be offered to a plant. Peaches naturally have a need for a long chilling requirement, a requirement that might be met in a place like Pennsylvania, New York, or New Jersey. Though peaches are naturally adapted to more northerly states there is little doubt that, chilling requirement aside, peach trees will grow more quickly, as most plants will, in a southern climate and so we have, over time, bred peaches to have shorter and shorter chilling requirements. These peaches aren't much different from their northern brethren except that they need less time in a cool environment to stimulate them to flower. Unfortunately for peach growers, this southern environment that can allow their trees to

grow faster and bring their product
to market faster also brings the
added threat of certain diseases and
insects which flourish in warmer
environments. In the 1800s and
early 1900s croplands across the
South had been destroyed by cotton and farmers needed something
to fill their fields. Peaches fit the
bill.

A lonely peach tree in the middle
of what was once an orchard.
Peach trees are short-lived, often
living less than 15 years, especially
in old orchards. This tree is about
20 years old. (*Jeff Gillman*)

Peaches developed into an
important Southern crop for three
reasons. The first is that constant
cotton cropping had robbed the
soil of nutrients and organic matter.
Cotton has been a southern crop
since the 1700s, and if it isn't managed well it will slowly turn
beautiful black soil into a layer of 6-8 inches of sandy loam over
red clay. Peaches are unique in that, with proper care, they can
live and sometimes even thrive on the sort of poor soil that cotton farming produces. Another important development for
peaches was the advent of the refrigerated rail car. Peaches do not
store well, and so in the early days of southern peach production
it was difficult to get peaches from the South to the North.
However, by the late 1800s refrigerated rail cars were being used
regularly and peaches could be shipped to Northern markets
with much less chance of injury despite the four to five days that
it would take them to reach their destination. The third and final
reason that peaches emerged as an important crop is the development of a particular variety of peach, the Elberta. Samuel
Rumph, a Georgia nurseryman, named this peach for his wife

and, though rarely planted today, Elberta is the most famous peach ever developed. It was the peach most frequently planted in the South until the 1960s. The Elberta's advantage was its firmness, which allowed it to be shipped more easily than its contemporaries. It didn't hurt that the Elberta was also large and flavorful, and could hit the northern market early in the season before northern peach trees could produce their fruit. Rumph was not just a grower, though; he is also one of the people who helped develop the refrigerated car for shipping the peaches. His development of Elberta and the method to ship it paid big dividends not only for the South, but also for himself.

As the peach industry grew across the South, certain areas became known as peach regions. Typically on old cotton land that was close to a railway, one of these regions sits smack in the center of Georgia in a small county called Peach. This county is not only in the middle of the peach state; it is in the middle of peach country, where the highest concentration of peaches in Georgia is grown. The peach trees that are planted here have a short, brutish life typical of peaches across the southeastern United States.

⊷

The tree was four feet tall when it arrived at the orchard. Its trunk was not a continuous straight line as it would have been had it come from a seed, but instead was crooked near the base. The crook was from a budding procedure performed to ensure that the delicate top of the plant would be connected to a root system that would be able to handle the diseases and other pests present in the ground that it was being planted into. A laborer placed the young tree into the earth at a point indicated on a string which lay on the ground with marks spray painted on it every fifteen feet. Running below the string was a narrow trench where he nestled its roots, firmly pressing soil over them with his boot. He quickly moved on to the next tree. Each tree planted earned the laborer another few pennies. He was paid by the work he did, not the time he spent doing it.

After it was planted, the little tree was watered into the ground along with its neighbors, each of which was of exactly the same size and structure. They had the same top and the same root system, which was designed to resist the attack of the many nematodes and diseases in the Georgia soil. Before these trees were planted, the earth had been sterilized. This sterilization meant that the tiny worms called nematodes which could injure the peach tree's young roots and might even kill it weren't in the soil right now, but they would come back. They always did. And when they came back, if the peaches weren't on the right root-stock, the young trees would be overtaken.

MUCH of the peach land in the Southeast has been used for this purpose for the past seventy years or more. With so many trees in the same place for so long, peach-hungry pests have become numerous in the soil. Unless you sterilize the ground using some particularly nasty chemicals and plant trees that can handle these pests, there's a good chance the trees will perish. Of course, there is the option of not planting peaches in the ground for a few years, which would also help to alleviate the problem, but that would mean a huge reduction in income that most orchardists can't afford. Besides sterilization there is a second method that orchardists use to control this problem: grafting. Like the grafted apples in the last chapter, the varieties on top of a peach tree will have wonderful names like Flamin Fury, Rich Rose, and Springprince, while the rootstocks have less attractive names such as Guardian Brand BY520-9. The reason for this discrepancy is the same as it is with apples. You buy peaches based on the fruit itself, not on the rootstock it's grown on.

Peach rootstocks are grown differently than apple rootstocks in a way that makes their production somewhat easier. Peaches can be selfed. Selfing means that a flower on a peach tree can fertilize itself. As with most fruits, apples cannot fertilize themselves. The inability to self-fertilize is an evolutionary adaptation, ensuring that offspring are genetically diverse. Peaches are odd because they can and do fertilize themselves, and for the orchardist this oddity is an advantage. Because peaches fertilize themselves, their seeds are very genetically similar to their parent, unlike the seeds of, say, apples, whose seeds contain genetic information from both parents. The similarity between mother trees and their offspring is not so small that the varieties of peaches used as scions can be grown by seed; however, they are close enough that rootstocks can be produced by seeds. These rootstocks are extremely important to peach production in much the same way that phylloxera-resistant American roots were important to European grape growers. Today's peach rootstocks are resistant to nematodes and a variety of other pests that devastate peach orchards. Unfortunately, with the problems that peaches face today, often resistant rootstocks just aren't enough.

Peach trees are susceptible to so many debilitating diseases and insects it's doubtful they could survive in the southern United States without human intervention. If all the peach trees in the South were suddenly abandoned and left to their own devices, I doubt 10 percent would be alive in ten years. As we will soon see, one of the most debilitating diseases of peach, peach tree short life, is also one of the most difficult to control. Occurring as a complex of bacteria, cold damage, and harmful nematodes, this problem can bring an orchard to its knees faster than almost anything. But this is really only the beginning. Insects have a role to play in the peach orchard too, particularly borers. For most trees, borers are secondary pests, attacking the tree only if it has already been damaged. In fact, borers often detect trees where they prefer to lay their eggs by sniffing the air for the compounds that are released by damaged trees. Peachtree borers are different, though. They are primary pests, attacking and often killing young healthy trees and eating deeply into the orchardist's bottom line.

Insects attack plants in many different ways. The most familiar attack comes from chewing insects because their damage looks the worst, but among pests, chewing insects are the least harmful insects out there. I don't mean they can't kill a tree, but they usually don't. Chewing insects include such beasts as tent caterpillars and Japanese beetles. They chew on leaves and reduce the tree's potential to make food for itself, but this isn't as big a problem as most people would think. The tree expects to be fed upon, so it produces more leaves than it actually needs and thus can lose about a third of its leaf area before it actually suffers. Seeing a tree with bare limbs makes us afraid for the tree, but for the tree itself, it's mostly cosmetic. A second way that insects can attack trees is by sucking sugar out of their leaves. Insects like aphids and leafhoppers feed by inserting their straw-like mouthparts into the leaf and sucking out the juices inside. These juices are full of the sugars made by the tree's leaves for the little bug to enjoy. In fact, these juices are usually so rich in carbohydrate and so poor in protein that the insect will end up eating a lot more carbohydrates than it needs to get to the protein. This results in the insect having a sugar overload, which is quickly remedied. Excess carbohydrates leave the insect in the form of honeydew (a sort of sugary poop), which is eaten by other creatures, most notably ants, who will often "tend" aphids like cattle, harvesting their waste. Boring is a third type of attack. If there's one insect that a tree doesn't want around it's a borer. Secondary borers, those which attack stressed trees, bring death more surely and rapidly than the stress itself would have, but primary borers go straight for the tree's heart, its vascular system, without waiting for stress to compromise the tree's ability to fight back.

The life of most borers starts on or just inside the bark of a tree where their mother lays an egg. The egg hatches and the small larvae start to eat, chewing their way into the tree's stem. If the tree is strong and conditions are right, the tree may be able to fend off the borer by "sapping it out," meaning the flow of sap inside its stem is strong enough to push the insect out of its tunnel, which will eventually fill with sap that hardens as it dries. If the tree is weaker than it should be, or if conditions aren't good for the tree, or if too many insects attack, or if the borer is able to handle the tree's sap defenses, then the borer won't be pushed out of the tree and will start to feed on the tree's cambium. This is the worst possible place that they could choose to feed on. When you peel the bark off a tree, the place where it separates from the wood is the cambium. New vascular cells are produced in the cambium and without it the tree will not be able to produce the tissues that it needs to carry water and nutrients up from the soil into the tree branches or carbohydrates down from the tree branches into the tree's roots. Trees can lose some vascular tissue without dying, otherwise every slip of the weed whip would be terminal. But if the borers damage the cambium in a ring around the tree's trunk, then the tree is destined for the

burn pile. Unfortunately, that is
often what happens. Even if an
entire ring isn't eaten away, the tree
is usually still stunted and sickly for
the rest of its shortened life.

In peach orchards, farmers
depend on chemical sprays to control borers. These sprays are applied
when the adults emerge from pupa,
which occurs at different times for
the different borers. The sprays kill

Peach tree borer, one of the most
feared peach pests of the southeastern United States. (*Mean and
Pinchy/Flikr*)

the adults trying to lay their eggs, and just about any other insect
in the vicinity too. If the farmer wants to use a less toxic method
of control they can apply a pheromone which mimics the
female's scent and confuses the borers while they're trying to
find mates. Without mates, there are no fertilized eggs. This
method isn't quite as surefire as applying pesticides, and so it isn't
the most popular option.

The peach tree borer, like grape phylloxera feeding on
European grapes, was never meant to feed on peach trees. This
borer is an insect that is native to the United States. Before
peaches reached our shores, it was a relatively minor pest of the
wild cherries and plums that thrived here before the Europeans
settled. By introducing the peach and growing so much of it,
humans brought this borer problem on ourselves. Another
scourge of peaches that is of our own making is peach tree short
life. And if one peach disease deserves to be called a plague, peach
tree short life is it, with as many as half a million peaches dying
annually of this condition in the Southeast.

Peach tree short life still isn't fully understood by scientists.
Many factors affect its onset, including pruning time, fertility,
cold, and water availability. Furthermore, various nematodes and

bacteria also play a part. What is particularly interesting about this disease is that it is, for the most part, a southern phenomenon. In other parts of the country it simply is not a serious problem. This disease usually begins as an infestation by ring nematode. This little worm is not considered a particularly damaging pest of most crops it can infest, which range from peppermint to carnations, and peanuts. However, when it infests peaches it predisposes the tree to cold damage and a certain type of bacteria, and when these hit the stem, the tree is as good as dead. There is currently no surefire way to protect a tree from peach tree short life, though orchardists have strategies to avoid the disease, such as avoiding early winter pruning, which seems to cause damage that may lead to this disease. The orchardist can also select rootstocks that are resistant to nematodes, the Guardian rootstocks being best, but even these precautions won't eliminate peach tree short life, just keep it to an acceptable level. The protection provided by this short life resistance is made even more misleading since Guardian isn't resistant to another disease, Armillaria, which attacks peach trees later in their lives than peach tree short life. Indeed, this rootstock now only seems to offer a temporary respite as Armillaria is becoming the dominant killing disease in the Southeast.

In the first and second year, our tree was pruned to encourage its side branches to grow and prosper, which resulted in a tree with an open center. This structure was developed so that sunlight could readily penetrate the canopy as the tree aged. Sunlight helps the peaches to develop their final colors. In the third year, the tree blossomed. The blossoms that the tree produced, however, were stripped off to encourage growth. It wouldn't be worth the orchardist's time later in the season to pick only a few peaches, and it wouldn't be a good idea to allow fruit to rot on the ground, potentially spreading disease to other fruits around the orchard. The wasted fruit was a shame, though; this tree produced June Gold, a yellow-fleshed peach that would ripen early in the season.

The little tree was steadily producing fruit now. Both of its neighbors had passed on, one to peach tree short life and one to peach tree borer. At five years old our tree was far from out of the woods, but with a constant stream of insecticides it was kept pest free. A length of hose ran along one side of the trees trunk, delivering water whenever the ground became too dry. Fertilizer was applied next to the tree twice a year to encourage its growth and the health of its fruits. Once a year it was pruned too. Its early training had resulted in the development of four main limbs growing out from the trunk about four feet off the ground. These limbs spread outward into a shallow canopy allowing sunlight to easily reach its ripening fruit. Most years the tree was pruned in the summer, but in this, its sixth year, pruning was done in November.

On the outside the tree appeared healthy and produced beautiful fruit. But the tree was on the edge of the orchard, and diseases were slowly starting to infiltrate the soil. Nematodes, deadly and almost impossible to control, were gaining a foothold.

The tree grew steadily thanks to regular inputs of water and fertilizers. Heavy doses of nitrogen, phosphorus, and potassium were added to the ground around the tree twice a year to replace the nutrients that left when the peaches were harvested. Nearby, no new trees had been planted where its old neighbors once stood; instead the spots were left open. The orchard was being

converted to a pick-your-own operation as it aged and the number of pesticides would be reduced to cut costs. This reduction
in pesticides would probably mean the eventual end of the
orchard though. Without pesticides the borers would gradually
destroy the remaining trees. This was just a way for the orchardist
to make a few more dollars on the orchard before it had to be
shut down and sterilized for another crop in a few years. Less
than eight years was a particularly short life for an orchard, but
that couldn't be helped. At the next planting more resistant rootstocks would be used and more care taken in the spraying and
pruning of the trees.

ORCHARDS are kept clear of weeds and other pests that might
interfere with the growth and fruit production of the trees
grown there. Because of that, in an orchard managed by people,
trees need fertilizer to thrive, whereas in the wild trees do well
without any additional help. In an orchard, people push trees to
the limits of their abilities and reap the benefits. In more natural
settings, trees produce fruit, seeds are in the fruit, and, when the
fruit ripens, it falls to the ground and rots. Animals, including
insects, may eat this fruit, or they may not, but whether or not it
gets passed through an alimentary canal, the nutrients in the fruit
make their way slowly back to the tree's roots to be recycled. The
limiting factor to the growth of any plant is usually the nutrient
that is least available to it, and that nutrient is usually nitrogen.
Over the years we have discovered that applying nitrogen to the
soil, along with other elements, is the best way to keep trees
growing.

In an orchard or farm, adding nitrogen is the key to getting
plants to grow quickly. In a more natural environment, such as a

forest, nitrogen is still the key to
faster growth, but here it is added
in a different way. In the forest,
nitrogen comes from bacteria
which take nitrogen right out of
the air and change it from a form
that plants can't use to a form that
they can use. These bacteria don't
replenish the soil with nutrients
as fast as fertilizers applied by
humans can, but they are remarkably effective and kept the plant
world nicely stocked with nitrogen until humans started getting
into the act.

A tree dying of peach tree short life
in a Georgia orchard. Short life is due
to a complex of many different problems and is among the leading causes
of peach tree death. (*Dan Horton*)

In a managed peach orchard, fruits are removed from the tree
and have no chance to be recycled for later use. Our treatment
of the ground, applying herbicides and sterilization before trees
are planted, prevents those bacteria that take nitrogen from the
air and make it useful to plants from working as effectively as
they should. Plants called legumes have bacteria living right in
their roots that can take nitrogen from the air and turn it into a
more useful form. These plants are sometimes placed in orchards
between rows of trees to add nitrogen back to the soil. But
legumes use the bacteria for their own needs and don't release
much back into the environment without the encouragement of
a lawnmower. The parts that are removed from the legumes fall
to the soil and the nutrients, especially nitrogen, that were stored
within are released back into the environment. Unfortunately, as
good as this system seems, it doesn't provide the peach trees with
all of the nitrogen they can use, and so orchardists who use this
system still often wind up adding more. Fruit production is the

key to a successful orchard and the orchardist will do anything to get the tree to produce more.

The idea of adding nutrients to plants to promote plant growth isn't new, though exactly what nutrients to use is relatively recent knowledge. The person most responsible for our current understanding of how to feed plants is Justus von Liebig, a nineteenth-century organic chemist. Among Liebig's most important contributions was his law of the minimum, which says that a plant is limited by the nutrient in shortest supply. This nutrient is usually nitrogen. Until the 1830s the most common way to get nitrogen to a plant was to apply manure, and most cultures have known for many years that applying manure or dead animal carcasses (which also contain lots of nitrogen) around their crops would encourage them to grow, though they didn't know exactly why. But with low concentrations of nitrogen these fertilizers were inefficient, and so, after people found out how important nitrogen was, they searched for a better source of this nutrient for crops. What was better than manure? Really, really old manure. In the early 1800s, about the time that scientists figured out how important nitrogen was for plant growth, they also figured out that there was a whole lot of nitrogen in guano. Guano is aged manure found in various places around the world, most notably bat caves (bat guano) and the shores of Peru, where arid conditions allow feces of birds who have been eating a diet very high in nitrogen (full of shellfish) to sit and age rather than being washed away. By mining this aged manure and applying it to land, farmers had an easy way to apply lots of nitrogen quickly, and guano was used with abandon. By the late 1800s Peru was nearly depleted of guano.

In the early 1900s Germany was experiencing a shortage of nitrogen. It wanted this resource for fertilizer, of course, but it also needed it for something a little less well meaning: bombs. Nitrogen is the primary ingredient for most explosives and, without a ready source, Germany would be in trouble if war ever broke out. And in the early 1900s, war seemed to be inevitable. Germany also knew that they would not be able to send ships unfettered to South America to grab what remained of the Peruvian guano. So the Germans put a great deal of effort into trying to get nitrogen from the most readily available source, the air. Because the nitrogen in air is very difficult to use to make fertilizers or explosives, it needed to be changed into a different form. Fritz Haber and Carl Bosch were the scientists who eventually worked out how to take nitrogen from the air, react it with natural gas at high temperatures and pressures, and make ammonia, which can then be used to make other fertilizers. This process uses tremendous amounts of energy, usually in the form of coal, to supply the heat and pressure, and also uses a tremendous amount of natural gas to react with the nitrogen. It is through this process that the nitrogen in most fertilizers is made today.

There are some growers of organic peaches, apples, and other crops who apply nitrogen from nonsynthetic natural sources to grow their food, but they still need to add nutrients to their soil to get the best yields possible. The originator of the organic movement, Sir Albert Howard, watched the Chinese grow crops by efficiently reusing plant and animal wastes (notably human wastes). Watching the reuse of these wastes inspired him to develop systems where wastes would be reused to promote plant health. This is the origin of modern organic growing. The fertilizers that organic peach growers use today may include compost, fish emulsion, cotton meal, or any of a variety of other natural substances that will offer nutrition to the crop.

✧

As our little tree entered its seventh year it no longer had a need for nitrogen, or any other fertilizer for that matter. Poor pruning and a rootstock that wasn't sufficiently resistant to nematodes had brought its life to an end. As spring growth flushed around it our little tree sprouted only from the base of its stem. Its vascular tissue was destroyed by a combination of cold damage and bacteria. When the bark around its trunk was cut by the orchardist to confirm the cause of death it emitted a vile sour odor, the most telling indicator of peach tree short life. Just below the damage new sprouts were beginning to grow, but these wouldn't be given the chance. This bit of the orchard had lost too many trees, and next year the land would need to be reworked. Seven years doesn't seem like a long time for an orchard to live, but it is becoming more and more typical. Even in the early 1900s orchards rarely lived beyond twenty years. Today that number is closer to fifteen. Because of how peaches have been cultured and shuffled from place to place over the years there is no way to know how long a peach might have lived if we had never domesticated it.

There were about 15,000 acres of peaches in Georgia, 18,000 acres of peaches in South Carolina, and 10,000 acres of peaches in Florida in 2006, and they're here to stay. Peaches are an institution in the southern United States. It seemed like such a good idea. But the peach was brought from a foreign shore, subjected to pests it should never have had to experience, and exposed to climates where it must be coddled and manipulated to grow. We should ask ourselves, exactly what we were thinking?

Besides the apples and peaches of the last two chapters, numerous other trees have been introduced to the United States. From Asia, South America, and Europe, plants came to feed, clothe, and provide entertainment for us. But trees and other plants certainly weren't the only things that came to our shores; other living things came as well. Some sneaked in and some were brought with a purpose in mind. These creatures have done as much as our introduced plants to alter the landscape around us, for good or bad.

Chapter Seven

FOREIGN INVASION

Thinking back to the yard where I grew up in Pennsylvania, it was a conglomeration of plants and animals from around the world: lilacs, apples, and peaches from Asia, grass and pines largely from the United States, dandelions from Europe, corn from South America out in the field. It wasn't just the plants though; insects and diseases from other regions dotted the landscape as well. Japanese beetles and codling moths, from Japan and Europe, respectively, combined with Colorado potato beetles to feed on the yard, creating a truly multicultural ecosystem: a real American melting pot.

So many diseases, insects, and other organisms have found their way across the seas that it is becoming difficult to tell what was here when settlers first came and what wasn't. The American earth itself is changing from these introductions. When glaciers covered the northern part of the United States they chased the earthworms out. Trees in this region lived in earth that was earthworm free from the time the glaciers receded until Europeans came. European species arrived in the ballast of their ships; these unintended passengers then made their home here after the ships emptied their ballast tanks. Most of us think of earthworms as good creatures, and often they are. In fields and gardens, they help to break down debris from dead plants and animals, and they keep the soil in gardens and fields well aerated. But earthworms aren't always beneficial. Faster decomposition of leaves on the floor of a hardwood forest creates compacted soil, which isn't hospitable to low-growing plants including young trees and ferns. Worms also loosen soil, making the ground more easily eroded and damaged by rains. Some trees, such as oaks, seem to prosper when earthworms are in the soil, but others, such as maples, do not. There is no reasonable way to get rid of the worms that we have introduced. Nor would most of us want to, as they perform some useful services for our gardens and farms. Because of their slow progression into the forests (earthworms can travel only a half mile over a hundred years) the consequences of their introduction into our northern forests aren't fully known and won't be realized for many years. The earthworm's slow changes to our forests creates a stark contrast to the more rapid changes brought about by other creatures.

The early settlers of North America came to escape oppression. When they arrived, they found a land of plenty where they could do what they wanted. There was enough for everybody and they were free from tyranny. And when you stop to think about it, isn't that just what any insect or plant disease wants? Lots of food, and freedom from its natural enemies? Like a rabbit suddenly released onto an island full of carrots without any cats. But their paradise is our problem. As these pests compete with us for natural resources we need to find a way to stop them, or live with less.

As a rule, most of the battles we fight with tiny invaders from foreign shores end in a stalemate. Neither we nor our insect competitors can claim a clear victory, but there are some notable exceptions. One of the great successes of American entomology is the eradication of the screwworm fly, a particularly nasty insect native to the tropical regions of North, Central, and South America. It was first noticed in the United States in the mid-1850s and quickly became a major problem in livestock production. This fly is a force to be reckoned with, a nightmare from a bad horror movie.

Adult screwworms are unremarkable flies which look something like oversized house flies with huge reddish-orange eyes. After the adult flies mate, females seek out a lesion on a mammal (like a cow, a horse, or, rarely, a human) in which to lay their eggs, and then things begin to get interesting. The screwworm gets its name from the feeding habits of its larvae. The larvae emerge and literally screw themselves into the host animal, feeding on their otherwise healthy flesh. Death (for the mammal) is often the final result. Before people started using pesticides to control them, screwworm flies could render a herd of cattle sick and useless very quickly. The flies were hated and feared by ranchers, and due to the abundance of beef farms, the fly's range grew quickly from a few isolated spots in the Southwest to most of the southern United States. In the early days of the screwworm many different tactics were used to control the flies, but few had much effect. When they became available, DDT and other powerful insecticides were applied to animals to protect them, but even these cure-alls just slowed the flies down. Complete prevention was impossible. Even if all the livestock in an area were coated in pesticides, it would be impossible to go into the woods and treat all the animals there that could play host to the pest. Eradication was just a dream.

Early in the development of modern insecticides, in the 1940s and 1950s, they seemed to be the solution to every insect problem. In fact, I have two editions of a book titled *Destructive and Useful Insects* in my bookshelves. This book was once considered the encyclopedia of insect control. In the second edition, published in 1939, all kinds of strategies are suggested for controlling various insects; from arsenic sprays to mounding soil around possible infestation sites and selecting proper rootstocks. In the third edition, published in 1951, almost every pest control method is a summary of how much DDT or some other insecticide to use. Over the course of only a dozen years these modern insecticides went from nonexistent to the most effective insect-control agents anyone had ever seen. They seemed as foolproof a cure for insect maladies as was possible. Most other techniques were ignored.

Because pesticides alone failed to control the screwworm fly, this insect was one of a few that forced people to admit that pesticides are only one possible tool among many. Control of this devil finally came about not through the use of pesticides, but rather through a careful analysis of the screwworm's sex life. It seems that the female screwworm only mates once in her life, and so, if she can be mated with a sterile male, she will not produce fertile eggs. But where to get sterile males? In the 1940s and early 1950s, research on radiation was all the rage in the scientific community. This research had shown us that careful application of radiation could destroy the reproductive abilities of many animals without killing them. Thinking this might help solve the beef industry's problems, a group of scientists tried this little trick with screwworm flies, and it worked magnificently. Sterile flies were first released in Florida in 1957, and the program worked so well that by 1959 the screwworm fly had been eradicated from the southeastern United States. About ten years later it had been eradicated from the whole country. But the fly

is still alive in other parts of the world and is accidentally imported into the United States every now and again. In fact, a case of
a screwworm fly on a horse was reported in Florida in 2000.

Somewhat less grotesque but equally scary to those who have
to deal with it is the boll weevil, another insect effectively eradicated from most of the United States. If screwworm eradication
was realized through careful analysis, then boll weevil eradication
was realized through brute force. The boll weevil was one of the
worst pests of cotton from the time of its introduction from
Mexico in 1892 until the 1970s when the eradication program
began. Cotton is not native to the United States but comes from
more southerly climates. Various species of cotton can be found
in more tropical regions around the world, including Central and
South America, Africa, and Asia, where cotton usually lives as a
perennial plant rather than an annual. The boll weevil, a native of
Central and South America, has always been one of cotton's
worst pests. The huge cotton fields that the United States had
developed in the South of the 1800s were practically begging for
an infestation, and the boll weevil was only too happy to oblige.

After the boll weevil first came to southern Texas it spread rapidly, sometimes expanding its range by sixty miles in a year. Efforts
were made early on to control it. The insect's life could be made
more difficult by destroying any plants left in the field in the fall.
Boll weevils spend the winter hibernating in these plants and so
their destruction limits the number of boll weevils that live
through the winter. Early crops were also encouraged as they
would miss the onslaught of boll weevils, which usually came
later in the season. As effective insecticides became available these
were encouraged too. Calcium arsenate was the insecticide of
choice until DDT came onto the scene. But these methods combined still only controlled a portion of the problem. Crop losses

were estimated to be somewhere
between 20 and 40 percent during
the first half of the twentieth century. Farmers were starting to lose too
much cotton and many were going
out of business.

With the advent of DDT and
modern pesticides in the 1940s, the
boll weevil's good fortune was chal-
lenged. DDT was effective against
this pest at first and knocked its
numbers down, saving many fields of
cotton. DDT used in combination

An adult boll weevil. These
insects are dependent upon cotton for their entire lives, making
them an easy pest to target.
(*Winfield Sterling*)

with the fall destruction of plant materials in the fields and
growing early crops worked even better. Unfortunately this success was short lived. By the 1950s the boll weevil had developed
resistance to DDT and many of its chemical relatives. The cotton industry needed something to get rid of the weevil once and
for all, and they got it. In 1971 a pilot program was initiated to
examine the feasibility of eradicating the boll weevil from the
South using what were then considered more modern pesticides,
mostly organophosphates. These poisons derived from research
conducted during World War II focused on creating nerve agents
for use against humans.

The boll weevil is dependent on cotton for most of its life,
feeding on the cotton's bolls (which are the seed pod of the cotton plant) as larvae and as an adult. With such a limited host
range there was little elegance to the eradication program for this
weevil, except for the development and use of pheromone traps,
which attracted males and helped to identify when the insect
was present so that farmers could accurately time their sprays. To
control the boll weevil, they simply waited until it was present in

a field and then applied pesticides with such incredible ferocity that eventually this pest was brought to its knees. This brutishness was possible because of the boll weevil's dependence on a single host plant. Unlike the screwworm fly, which has many other possible hosts, the boll weevil needed cotton and could not survive on other plants. There is little wild cotton in the United States, and so when growers apply pesticides to cotton fields, they are poisoning the only food the boll weevil knows. It has nowhere to hide.

The boll weevil has been effectively eradicated from a long list of states, including Virginia, North Carolina, Georgia, Florida, Alabama, Kansas, Arizona, New Mexico, and California, but one will still show up from time to time, and if we're not careful this insect could come back to haunt us. A short time of intensive spraying led to a long period when fewer pesticides needed to be applied, but farmers need to keep their guard up or something else could slip in and exploit this rich resource. Despite the theoretical reduction in spraying that boll weevil eradication was supposed to bring about, the presence of boll worm, a moth whose larvae prefers to feed on the bolls of cotton, and stink bugs, who insert their long mouths into cotton bolls to feed, have conspired to make the victory bittersweet.

But for all the successes with insect eradication, there are also many failures. The bean weevil, the elm leaf beetle, and the codling moth have all come to the United States from across the sea and successfully established themselves. These insects and others like them have traveled here in a variety of ways. The first insect pests to arrive here traveled in ship's ballast, like the earthworm, but other insects traveled as stowaways on stored grain, food, or people. Seed-eating weevils, cockroaches, and bedbugs were some of the first insects to find their way onto American shores, but they were far from the last.

WHEN I was eight years old or so I have a vivid memory of going out to the front lawn and clutching a piece of turf. As I pulled upward I remember being surprised at how easily it lifted from the soil, almost like a throw-rug. Underneath this section of grass sat five or six small beetle larvae curled into the shape of a C. This was my introduction to the Japanese beetle, one of the most voracious eaters ever introduced to this country. Besides the damage that its larvae inflict on grass roots, the adult beetle chews the leaves of a wide variety of shrubs and trees. From plums to peaches and roses, I had the opportunity to see the unsightly mess this beetle leaves wherever it landed.

Introduced to the United States around 1916, the Japanese beetle was quickly identified as a bad player. No one knows exactly how the Japanese beetle came here, but it seems likely that it was a stowaway on food materials arriving from Asia. In Japan, this beetle is commonly seen but is not considered a major pest. In the United States things were different. The beetle was first found in New Jersey, and caused such damage that by 1917 the Department of Agriculture had begun an investigation into the detrimental effects the beetle would likely have. They soon established that the eastern United States was a perfect place for the beetle to grow and prosper. It quickly spread from New Jersey across most of the Northeast leaving skeletonized leaves and throw-rug lawns in its wake.

By 1918 the Department of Agriculture had decided to try to eradicate the beetle, but funding was scarce and the Japanese beetle population turned out to be bigger than anyone had at first thought. The beetle had too much of a head start for initial efforts to be very effective. Compounding this problem was the fact that most of the pesticides of the day just didn't work that well against this beetle. DDT's insecticidal properties hadn't been identified yet and organophosphates wouldn't be available until World War II. To protect fields of turf, recommendations were for 435 pounds of lead arsenate to be spread on the ground per acre. This treatment is an obvious case of the cure being worse than the disease. Soon after the Japanese beetle's introduction, efforts were undertaken to find natural enemies that could be imported to kill the beetle, but even after decades of searching, few are viable. The most effective natural control today is a disease native to the United States, milky spore disease, a bacterium originally found in New Jersey in 1933. Though this disease likes to infest beetles, it hasn't yet shown itself capable of eradicating this pest. By the 1940s, when DDT was first used against the Japanese beetle, applying pesticide to the area that needed to be covered for eradication simply wasn't possible despite the fact that it was an effective Japanese beetle poison. We have made other discoveries and invested in other means of controls such as the Japanese beetle traps that you can find in your local garden center. Unfortunately, these traps attract more beetles than they catch. If you have the Japanese beetle in your trees you have two choices, let them be or take solace in the temporary reprieve offered by insecticides.

Today, the Japanese beetle is not ubiquitous in the United States, being limited largely to the East Coast, for one reason: we are careful not to transport it. Ever since this beetle was identified, the U.S. government has used inspections and quarantines to limit its movements. Shipments of food and plants that are moved to the West are inspected to restrict beetles from making their way there. But the Japanese beetle is good at traveling, and even if it is never able to make its way across the Great Plains (where foods that the Japanese beetle feeds on are more scarce) in a car, truck, or airplane, which seems unlikely, it is all but a

foregone conclusion that it will still
eventually get there on its own six
legs and two wings.

Insects are only the tip of the
iceberg when it comes to plant
pests from foreign ports. Americans
have also brought a number of diseases to our own doorstep through
lack of care, or lack of foresight.
One of the first was wheat stem
rust. Wheat was introduced to the
Americas in the 1620s, and wheat
stem rust quickly followed, around

A Japanese beetle. These insects
feed on the foliage of plants across
the eastern half of the United
States. Their larvae are a pest of
grass roots. (*Jeff Hahn*)

1660. This fungus traveled here not on wheat but on barberries,
which are known for their beauty rather than their fruit.
Europeans brought barberries to the United States with the
intention of creating a homier atmosphere, and in the process
they delivered a major pest to one of the principal crops.
Scientists at the time had no way to prove that this disease of barberries also affected wheat, but their observations that the two
went hand in hand led to the enactment of a law in 1726 requir-
ing the eradication of barberries in Connecticut.

To prevent foreign diseases from affecting food, government
scientists in groups including APHIS (Animal and Plant Health
Inspection Service) spend a considerable amount of time figur-
ing out which foreign diseases pose risks to our crops and how
to deal with them should they ever infest the United States.
Plum pox, for example, is a disease of fruits including peaches
and plums. First identified in Bulgaria in the early 1900s, this disease quickly spread throughout Europe and had a pronounced
effect on crops. Plum pox is a virus that manifests as mottled

leaves and ring patterns on fruit and, though it doesn't usually kill a tree, it can halt fruit production.

One of the things that make plum pox particularly insidious is that a tree can be infected with the disease for three years before it shows any symptoms. Indeed, plum pox has entered the United States on a few occasions, and due to its ability to hide, it is possible that the disease was spread before we even realized it was here. This makes its discovery in Pennsylvania in 1999, Canada in 2000, and New York and Michigan in 2006 of concern. As soon as a tree in a region is identified as having the plum pox virus, that geographic area is prevented from shipping fruit to avoid further infestations, and any trees suspected of having the disease are removed, usually with a bulldozer. Plum pox is not a disease that can be cured with the usual array of pesticides, and so it is controlled through the death of its host, the tree in which it is living.

Where the plum pox carousel stops nobody knows. It sometimes seems like a game. How long can we keep this disease out of our fields? What can we do to eliminate it if it does get here? What if we can't eradicate it? But plum pox is only the beginning of the story. With plum pox we know what to look for and our orchardists and public servants will react as soon as they see a problem, hopefully nipping it quickly in the bud. But there are so many insects and diseases which could cause problems that we simply aren't aware of. Who could have expected that the Japanese beetle, such a minor pest in Japan, could become such a major issue in the United States? And if we're not looking for them, how much damage will be caused before we figure out they're here?

So many insects, so many diseases, so much devastation. Humans deliver them from here to there usually without having a clue that we're doing it. Sometimes I think humans treat the world like a blender, indiscriminately plopping things in: peaches and blueberries, ice cubes, maybe some rum and then choose a speed and set it spinning. When the blender is first turned on and for a few minutes thereafter, you can distinguish the different things spinning around behind the glass, but over time they start to mix together until, finally, it's a homogenized mess. The blender has been turned on, and the ingredients, which include all the animal and plant species in the world, are starting to homogenize. We introduce new things that live in, feed on, and otherwise damage the plants that we hold dear, changing the United States's ecosystem permanently for better or for worse. No tree is safe from our inadvertent introductions. Perhaps the best we can hope for is to keep the blender a few notches short of puree. Unfortunately for some trees, it's already too late for that.

Chapter Eight

THE PLAGUE
OF THE ELMS

ALL too often, a disease rampages through a human population, destroying lives and bringing tragedy. Once upon a time, such outbreaks would go unchecked, but today, diseases are contained by technologies of detection and treatment, and even potentially horrific killers such as HIV can be held at bay by education and drugs. Antibiotics, quarantines, and vaccines have all helped to limit the spread of these killers, and so major plagues have become rarer as time goes on, but before modern medicine, plagues occurred much more frequently. In the past, every hundred years or so a plague would rip through Europe and severely deplete the population, the most famous being the black plague, which moved from Central Asia into Europe in the mid-fourteenth-century. This plague was introduced to Europe by Italian ships escaping the trading city of Kaffa (a city in the northern portion of the Black Sea) where the Italians who controlled this city had been besieged by Tartar forces. The Tartars' siege ended earlier than they would have wished because of the onset of a strange and deadly illness. Before they quit their campaign, however, the Tartars hurled disease-ridden bodies into Kaffa with catapults, thus spreading the disease further. Ships that left Kaffa to escape to Italy came into the ports of Genoa and Venice with their crew members infected and often close to death. The wise thing to do might have been to cast the ships off and leave them to their fate, but instead they were allowed to dock and the plague came ashore. The disease soon spread throughout the continent, leaving whole towns empty in its wake. Known as the black plague for the dark appearance of its victims, this disease killed somewhere between one-third and two-thirds of Europe's population.

One of the remarkable things about plagues is that humans spread these diseases despite themselves. Were it not for a fascination with foreign goods, there would have been no reason for an Italian ship to sail to the East. Without this journey the plague might have remained there instead of spreading. If the people at the dock hadn't desired the trade goods, perhaps the ship never would have docked. If there had been less travel from Italy to other areas of Europe, the disease might have stayed in that country. But that is not the way we work. Humans like to have more and better things. We are a greedy species and, as our methods of transportation have become more efficient, we have become better at moving things: goods, money, and, regrettably, disease. Though HIV has only been known since the 1980s it has spread across the entire globe and killed over 25 million people, roughly the same number killed by the black plague.

We tend to think of plagues as being confined to our species without considering other creatures, but a quick check of diseases affecting other animals tells a different story. Cat plague, distemper, and mad cow disease are all plagues affecting animals other than ourselves and whose spread is encouraged by humans. And what about plants? There are a number of plagues that can and do affect the life of trees, and we have encouraged their spread. Over the past century and a half, no less than twenty potentially forest-devastating organisms have been released into the United States. Many of these outbreaks have reached plague status, including American chestnut blight, sudden oak death, laurel wilt, beech bark disease, and, most infamously, Dutch elm disease.

There was a time when an American street wasn't a street unless it was lined with elms. They were the mainstay of the urban and suburban landscape, their long arching branches shading streets and houses across the country. Fast growing but long lived, strong, tolerant of bad soils, and evenly branched, elms could rapidly provide a street with shade and, perhaps more important, character. The American elm, practically a symbol of the United States, along with rock and red elm, was the most planted tree in the late 1800s and remained so until the mid-1900s, but the planted elms are only one small piece of the elm story. Elms were once found across essentially all of the eastern and middle portions of the United States, from Maine down to Florida and west into Texas and through the Dakotas. They were rarely found in pure stands, but rather shared the forest canopy with other species including silver maple, sweet gum, sycamore, green ash, and others. Elms were prized by settlers for lumber, shade, and firewood, but they are fragile when outside of their native range. A number of diseases and insects attack these trees in Britain and much of the rest of Europe, and so American species are not particularly well suited to those and certain other areas of the world. Besides, Europe has its share of native elms, as does Asia. Native American elms are far from the only elms out

An elm-lined street in Green Bay, Wisconsin, in the early 1930s. This street is typical of the way streets looked across the United States before Dutch elm disease began to take hold. (*Wisconsin Department of Natural Resources*)

there. The elms are a very diverse group of plants, with many species, and chances are if one species of elm doesn't quite fit a location, another will.

We Americans have an odd propensity for overdoing things, and when we identified the elm as a high quality tree we found ourselves unable to stop planting it. After all, if a little of something is good, then a lot must be better. So it was with elms. Identified early in American history as the perfect street tree, these beauties were planted without regard for future problems and soon formed a canopy over streets across the United States. The elm we're going to look at in this chapter doesn't come from one of these planted trees, though; it comes from a native stand of American elms. One of many at the time the seed germinated, and one of just a few left to us today.

꧁

A seed fell from the elm, circling slowly as the wind caught it and tossed it through the outstretched limbs of other elms, lindens, and ashes to finally rest on a rich bed of forest soil. The forest was dense, but not overwhelming, along the edge of a man-made clearing. Only a short distance away, a farmhouse and barn were next to a wide lake. The seed sat through the winter, and from it a young tree emerged in the spring. The seedling grew at a slight angle, reaching for the light that filtered through the edge of the clearing.

This little tree sat in a corner of Kandiyohi County, in the middle of Minnesota. In this county sits a small town, little more that a dot on a map. Once upon a time this place was destined for larger things. It was slated to be the capitol of Minnesota in the 1850s, but the Dakota war of 1862 ended all of that. Only six miles from Kandiyohi, thirty-eight Sioux were hanged in what is still the largest mass execution in the history of the United States. Fearing further turbulence in the area, officials located the capital of Minnesota in St. Paul, an established city that was secure from possible American Indian uprisings. Today, the little town of Kandiyohi is a quiet place. Suburbs threaten, but are not a foregone conclusion. Houses are spaced far apart, and inhabitants are pleasant and unhurried.

At the same time this seed was germinating, a plague was brewing in northwestern Europe. World War I had ended and Europe was rebuilding, and replanting. The war had seen a tremendous loss of life. Though the most tragic were certainly the human lives that were lost, many trees were destroyed as well. Huge areas were cleared to provide lines of fire for machine guns, tanks were introduced that could turn a small tree to matchsticks, and massive artillery that cut through forests in search of softer targets was used on a larger scale than ever before. After the war, forest trees regenerated mostly without help, but in the towns of Europe people were seeking normalcy and planted trees wherever they could. But the planting was not necessarily going well. In France, Holland, Belgium, and Germany, a new disease had been discovered that was attacking elm trees. The first records of this pathogen appeared in those countries in 1918, 1919, 1919, and 1921, respectively. As it spread, plant pathologists struggled to figure out what, exactly, the disease was. The science of plant pathology was just coming into its own, and the study of harmful bacteria and fungi was in its early stages. The fungus causing the disease was first identified by the pioneering young plant pathologist Marie Beatrice Schwarz in Holland, which resulted in its name, Dutch elm disease. The spread of this disease, which reached England by 1927, seems pedestrian when compared to the spread of a human plague, but its pace was startlingly fast for a plant disease. The reason for the rapid spread was soon discovered, when an insect that could transfer the pest from one plant to another was identified.

In 1905, the first record of a breeding population of European elm bark beetles in the United States was made, in the New York City area. This record might well have been unimportant under normal circumstances. The European elm bark beetle is not particularly damaging to elms, so long as the elms are healthy. The beetles prefer to attack stressed and even dead trees rather than healthy specimens. In fact, if not for Dutch elm disease, we would probably hardly notice these little pests today, especially

since we have our own native elm bark beetle (which is very
similar to the European variety except that it isn't quite as good
at transferring Dutch elm disease). Unfortunately for us, the life
cycle of the European beetle ties neatly in with that of Dutch
elm disease.

In the early 1900s, people in the United States were just getting
used to the idea of tree plagues. In particular we were reeling
from the spread of chestnut blight. If the spread of Dutch elm
disease across Europe was fast, then the spread of chestnut blight
across the United States was like greased lightning. The
American chestnut was once a prominent tree on the East Coast
of the United States, with some forests composed of as much as
25 percent chestnut. It was a magnificent tree, often reaching 100
feet in height and with a trunk that could reach eight feet across.
Its wood was extremely strong, making it an attractive choice for
lumber throughout the eighteenth and nineteenth centuries. The
twentieth century would reveal a tragic fate.

In 1904, curators at the Bronx Zoo in New York identified a
problem with their chestnuts. A disease infiltrated the bark of the
tree, destroying the vascular tissue and slowly strangling the tree
to death as the fungus worked its way around the circumference
of the trunk. The damage makes the bark of the chestnut, normally a beautifully smooth-barked plant, ugly and hard to miss.
I can only imagine what those first observers of this disease
thought when they first saw it, but they were far from the last to
observe this damage. By 1908, this blight had affected chestnuts
across New York, New Jersey, Virginia, and Maryland. Identified
as a fungus that attacked chestnuts in the Far East, the blight led

to quarantines to prevent its spread. Chemicals were also tested, but none had any measurable effect. Quarantines and fungicides are laughably ineffective with a disease like chestnut blight, which we now know can be transferred by insects, animals, and even the wind. Even chopping down all of the chestnuts in an area can't stop this pest because it can live on oaks, too, though it won't damage them nearly as much as chestnuts. There are too many places for the disease to hide, and too many routes of transportation for it to follow for people to have any chance at containing it.

Chestnut blight showing the cracking and fruiting fungus on the tree's stem. (*USDA*)

No one is sure exactly how chestnut blight made its way to the United States, but it seems likely that it was imported through a ship's cargo, perhaps even as early as ten years prior to when it was first identified in New York. Within fifty years of its appearance in New York, there were almost no American chestnut trees left. However, because the disease affects the trunk of the plant rather than the crown and roots, fallen trees resprout, and so, although we lost the chestnut tree, we gained the chestnut shrub, a sad substitute for a once proud species.

As chestnut trees were destroyed, various tactics to retard its devastation were used and discarded. One tactic in particular was held onto by chestnut enthusiasts with grim determination: breeding chestnut trees that were resistant to chestnut blight.

Chestnuts across the United States that showed any resistance to the disease were propagated and grown in the hopes that they could help rejuvenate American chestnut forests, but all of this breeding ended in failure. By 1960, because of lack of success, the USDA pulled funding from blight-resistant chestnut breeding programs. Other groups never gave up. Even today, the American Chestnut Foundation crossbreeds American chestnuts with chestnuts from other parts of the world in hopes of creating a resistant tree, and the American Chestnut Cooperators Foundation attempts something similar but breeds only native chestnuts with hints of blight resistance. There have been successes, and today there are American chestnut trees with varying degrees of resistance, but, as yet, there are no plans for large-scale reintroduction of this plant back into the wild.

As our elm grew alongside other elms, most of its energy was directed upward, toward a nearby opening in the forest canopy. Its lower branches spread out into the clearing to catch the sun's rays, blocking them from other, smaller trees, allowing it to out-compete its nearby rivals who wanted the light for themselves. Deeper in the forest, other young elms were fighting a harder battle. The greater shade of the deep forest meant that they must grow straight upward to find any light at all. Lateral branches were smaller and less significant. At the top of the canopy, mature red elms displayed wild corkscrew branches that raced off in all directions, attempting to cover every inch of sky. The bottom two-thirds of the trunks were almost devoid of lateral branches.

At the age of fifteen, the tree matured, producing seeds for the first time. Now taller than any of the nearby shrubs, it had an expanding canopy that excluded all but the most aggressive understory plants. Over a thousand miles away, in and around the New York and Boston areas, its cousins were starting to die.

THE first reports of Dutch elm disease in the United States appeared in the late 1920s, the fungi having come across the Atlantic in lumber from Britain. In 1930, a population of trees with Dutch elm disease was found in Cleveland, but the disease remained localized and did not spread to other areas. Localized populations were also found in Cincinnati, Indianapolis, Norfolk, and Baltimore. By 1940, a 5,500 square mile area in and around New York City was infested with Dutch elm disease despite efforts to stop it. The reason for the contrast between the rapidly spreading epidemic in New York and the containment of the disease in other locales was easy to surmise: a population of European elm bark beetles already existed in the New York area.

The European elm bark beetle, and its relative the American elm bark beetle, feed on twigs of healthy elm trees as adults. Their feeding is usually concentrated in the crotch area between a young limb and the plant's stem. In extreme cases, I've seen twigs snap off from the trunk, but this is unusual and, by itself, is rarely enough to seriously damage a tree. After the adult elm bark beetle eats and, in due course, mates, it will need to select a site to lay its eggs. These beetles are rather picky and will not lay eggs on healthy trees, preferring to deposit their offspring on the trunks of damaged elms. The eggs hatch into small grub-like larvae which burrow into the space between the tree's wood and bark and feeds until it pupates. After pupating, the adults emerge from the tree and fly off to find suitable elm branches to feed

American elm bark beetle feeding at the crotch of a limb. These beetles transfer Dutch elm disease from tree to tree, ensuring that the disease is spread efficiently. (*Department of Entomological Sciences, University of California*)

upon, thus completing the cycle. Bark beetles and Dutch elm disease are perfect partners in crime. When an adult elm bark beetle emerges from a log killed by Dutch elm disease it has fungal spores on its body and so spreads the disease to healthy trees. When the fungus is introduced to the trees it will infect them if it can, weakening the tree severely and often killing it. This chain of events is fortuitous for the elm bark beetle, who can then easily find a conveniently weakened or dying elm log on which to lay its eggs, and so the cycle continues.

When Dutch elm disease first came to the United States, efforts were made to slow its progress. Unlike chestnut blight it looked like there might be a chance to stop it. After all, if it weren't for the elm bark beetle this disease would hardly spread at all. At first, eradication efforts consisted of destroying infected elms quickly, but as modern pesticides became available they were used aggressively to control the elm bark beetle, often with good results. The chemical used to control the elm bark beetle was usually DDT, which was applied over large areas. This method of controlling the beetles seemed to work relatively well, though it had some obvious drawbacks for the environment. In 1952 Curtis May and Whiteford Baker concluded in the USDA's *Insects: The Yearbook of Agriculture*: "If recommended spraying and sanitary measures are combined, it is likely that little loss from disease will result." Little did people realize at that time what they were seeing was just the tip of the iceberg. And

Dutch elm disease was a *Titanic*-sinking iceberg, a disease that would eventually take the life of our elm in Minnesota along with many, many others.

The elm tree's canopy was almost as tall as any of the others in the forest, and its branches and leaves spread well into the clearing. Birds nested in its limbs, and a variety of caterpillars fed on its leaves. The farmhouse was full of people who cared for their nearby crops and never gave a second thought to the forest, except as a source of firewood or sometimes lumber. This particular section of land had never been harvested for trees, unlike so many nearby regions. No, this land had been kept free of the forester's axe, and the farmer intended to keep it that way. He liked his forest as it was.

The transportion of diseases from one tree to the next by insects is common. In many cases the disease and insect are dependent upon each other, as with Dutch elm disease and elm bark beetle. Sometimes a disease will just tag along with an insect as it flies from tree to tree, as with chestnut blight. Either way, the disease has found a convenient means for spreading itself from one host to the next. In the same way, numerous human plagues use parasites for transportation: yellow fever, bubonic plague, and Chaga's disease. Phony peach disease, a bacterium that infects peach trees and also causes Pierce's disease, which affects grapes, are both examples of infections that are transmitted with the help of an insect. One of the benefits for a disease that uses an insect as a means of transmission is that the insect will often eat through the protective bark or epidermis around a trunk or leaf and so will circumvent the plant's defenses, giving the disease a free ride right into the most susceptible part of the tree. Quite a deal. Pierce's disease was a substantial but manageable problem before it was transmitted by insects; after that, it became a real scourge.

Many diseases and insects can be traced back to the point from whence they originated, but Pierce's disease is a little different. No one knows exactly where Pierce's disease came from, but it was first found in Anaheim, California, in 1892. This disease kills its host, grapes, by clogging the xylem of the vine and preventing nutrients from reaching the leaves. Within fifteen years of its discovery, the bacteria had destroyed nearly 40,000 acres of grapes in California. It was quickly established that the best way to control Pierce's disease was to plant resistant varieties, get rid of infested plants, and hope for the best. This strategy worked remarkably well for about ninety years. In the early 1990s, however, an insect pest was introduced that facilitated the spread of Pierce's disease to an almost unmanageable level. The glassy-winged sharpshooter, an insect which originally comes from the southeastern United States, found its way to the California coast, where it was only too happy to enjoy sunshine and fine wine grapes. Pierce's disease and the glassy-winged sharpshooter got along famously. The glassy-winged sharpshooter acted as a convenient set of wheels that Pierce's disease used to travel from vineyard to vineyard to wreak havoc. Without an effective control for Pierce's disease, many plants succumbed to this pest. The best way to keep the disease under control was to control the insect, but this was not an easy task. It takes flight quickly and is scared easily, so when the tractor carrying the pesticides sweeps by, this little insect is often long gone. When it sits on a leaf it sits on its tiptoes without much of its body touching the leaf, so even a heavy dose of poison isn't nearly as effective as it should be. Making the problem worse is the fact that this pest feeds by plunging its straw-like mouthparts into the leaf, effectively avoiding contact with the outer portion of the leaf where the pesticides sit. All in all this makes for a tricky situation: a disease that we can't cure transmitted by an insect that's tough to kill. Today, the best we can do is apply insecticides, remove infected vines, plant resistant grape varieties, and use a little bit of prayer. It's really not that different from what people who are confronted with Dutch elm disease do.

Dutch elm disease is a vascular wilt, meaning that it attacks the vascular tissue of the plant. Take a branch from almost any growing tree and peel off the bark; it comes off in a sheet. Under the sheet is the wood of the tree which will feel wet to the touch, particularly in the spring and fall when the tree is growing. This is where the tree expands and girth is created. Growth from this area pushes the bark of the tree outward, and so increases the diameter of the tree. This growth also forms new vascular tissue as, over time, old vascular tissue becomes unable to conduct water and nutrients. Any disease that attacks this area is very serious and can threaten a tree's life, and Dutch elm disease is no exception. What's worse, Dutch elm disease isn't really just one disease.

Just as there are a wide variety of dogs—pit bulls and dachshunds, poodles and greyhounds—there are also different varieties of most diseases. And, much as you'd rather be cornered by a shih tzu than a Doberman, there are certain strains of diseases that are significantly worse than others. In the case of Dutch elm disease, the first strain reported by the Dutch, the strain that crossed

Europe and later entered the United States, was not devastating. It was serious and it caused many trees to die, but it was not the kiss of death, and trees did recover from it. In Europe, where the disease began, it was common for up to 50 percent of an elm population to be killed in an area, but rarely was the area left devoid of elms. England seems to have had it best, for a time. There the disease took hold more slowly and trees were more likely to recover.

In the late 1960s, the ability of England's trees to avoid severe outbreaks came to an abrupt end. In the southern portion of the country, large numbers of elms started to die due to what was apparently Dutch elm disease. By 1976, England had lost an estimated 9 million elms out of a total of 23 million. It was soon discovered that this loss was due, not to the old strain of Dutch elm disease, but instead to a new, more virulent strain which could be traced back to a shipment of logs that came from Canada in the mid-1960s. This new, aggressive strain, the strain that we are now most concerned with here in the United States and the one that was so devastating in England, is known as "the aggressive strain." The older strain is known as "the nonaggressive strain". The aggressive strain is suspected to have been in the United States as early as the late 1940s. By the 1970s it had destroyed almost all native American elms. In the 1990s, a number of different programs began to see results from their work to find resistance, and a number of new varieties of American elms suddenly became available in the marketplace, including the now popular Princeton elm (which is actually an old cultivar that found new life when it was discovered to be resistant to Dutch elm disease) and Valley Forge elm. A number of other elms were introduced that were crosses between various Asian and European elms, many of which have performed very well. But resistance is not immunity, and many of these trees can develop Dutch elm disease and will if under stress. Many of these trees come down with and even lose branches to Dutch elm disease, but are ultimately able to survive the infection by limiting the disease's spread to a few branches. Unfortunately that wasn't the case with our ill-fated elm.

In 1968, a pathologist bought the piece of land in Kandiyohi where our elm tree sat. By this time the tree was reaching its full height of a hundred feet and becoming a dominant tree in the forest canopy. The normal lifespan of an elm that has reached the canopy is well over a hundred years, but when a disease enters the picture all bets are off.

High in the tree's upper branches, beetles ate in the spaces between the twigs and trunk, sometimes doing so much damage that a bud would fall right off. But this small loss didn't affect the tree much; it kept right on growing toward the sun, stretching to outcompete its neighbors and expanding its limbs into spots where the canopy faded as neighboring trees fell.

Then, one day years later, the pathologist walked out of his house in the late spring to see that the limbs of one of his favorite elms had no leaves. This was the first sign that the tree had contracted Dutch elm disease. He knew what needed to be done, and within a week he had girdled the tree to kill it. Soon afterward he cut the tree down to use as firewood. In the forest there were other, healthy elms. This was one of the last large stands of American red and rock elms in the United States and he intended to protect them for as long as he could.

Because of Americans' infatuation with a single type of tree during the end of the nineteenth and the beginning of the twentieth centuries, a magnificent plant was lost. By planting elms everywhere and introducing a convenient vehicle, humans provided a means by which Dutch elm disease could jump from tree to tree across miles and miles in any direction. And then, even though the disease threatened, people continued to plant elms, right up until there was no way around the problem. No tree should be planted so heavily that a disease can spread like Dutch elm disease did. There needs to be many species and varieties of trees planted in any environment, to slow the spread of any ailment that appears. We can never take control over a disease for granted. They can change over time, and we need to be prepared.

Chapter Nine

UNSTOPPABLE
INSECTS

Dutch elm disease could never have reached the proportions that it did without the aid of an insect. And the insect could never have made it to the place where it did without the help of humans. If people had had some foresight and planted elms more sparsely across urban and suburban landscapes, things might have gone better when Dutch elm disease struck. The habitual practice of planting one type of plant over vast areas is unbelievably attractive to insects that happen to eat that plant. Like a starving man who is offered a huge table of chocolate cake or cookies, insects will dig into these plants with abandon, leaving as few scraps as they can. And since over 99 percent of the cropland in the United States is used for growing crops that come from somewhere else, it's no surprise that, if an immigrating pest can't find a plant native to the United States that it likes, it can dig into something that it's more familiar with. And if it's really lucky it will find its food source completely unprotected.

Over the last five hundred years, more than two thousand insects from somewhere else have established themselves in the

United States. Some of these insects are helpful, such as certain introduced lady beetles and honey bees, but most aren't, and some are particularly noxious because of the broad range of plants they are capable of feeding on. There are lists of insects that the U.S. government officials are particularly concerned about because the pests have the potential to attack crops in this country and don't have any known predators here. This list includes such insects as the European chestnut weevil, the European cherry fruit fly, and the European spruce saw fly, whose preferred foods are relatively obvious from their descriptive names. All of these pests are potentially destructive to the crops they feed on, but because of their dietary restrictions we know where to watch for them. But there are plenty of insects that humans have brought to North America that have an eclectic diet that consists of just about anything green. These insects are hardest to control because they can hide just about anywhere.

So often when we think of plagues we think of diseases. When we think of human plagues we think of the bacteria that causes the plague and maybe the rats that carried it. When we think of plagues affecting plants we usually think about fungi such as the type that caused the Irish potato famine. When we think of an insect involved in a plague we usually think of it as transmitting the disease rather than being the problem. To humans, the mosquito carries yellow fever, and the flea carries bubonic plague. To trees, elm bark beetles carry Dutch elm disease. But insects may themselves be a plague. Like the biblical plague of locusts that decimated Egyptian crops, the phylloxera that destroyed the European wine industry, and the Colorado potato beetle that deeply damaged the potatoes of the West, there are insects that on their own can destroy whole populations of trees.

A host to feed on is the most important factor in the spread of plagues. Without a host the pest can't survive. Just as Dutch elm disease couldn't have spread if there were no elms around for the disease to live on, insects need to have a jumping-off point too. We tend to have a plague of insects when two conditions are met. The first is that we must have something planted with enough density that the pest can move easily from one plant to another. The second is that there are few or no natural enemies prepared to eat or parasitize this new pest. (It's never a surprise when a new pest comes over from another continent and has no enemies. After all, why would it? It hasn't had enough time to develop any yet.) Given the way humans grow certain plants while completely ignoring others, it's no surprise when an insect reaches plague status, and so it's also no surprise that a young ash meets its eventual fate because of overplanting and a careless introduction.

The tree was magnificent. It was planted in a huge field with thousands of other ash, but it was the best of the bunch. At this stage of its life there was no discernable crook from the connection between living tissues where the bud from a Patmore ash had been placed onto the seedling rootstock at the nursery. After five years of living in the ground the tree had done well and would command a very reasonable price when it was sold. It was lifted from its spot in the nursery with a mechanical tree spade. The diameter of the soil ball in which its roots rested was over forty inches and weighed close to four hundred pounds. The soil ball would be wrapped in burlap and wire as soon as the rest of the harvesting crew caught up. It was a four-man operation, one person operating the skid-steer loader on which the tree spade was mounted, one person placing the basket and burlap so the tree could be placed into it, and two other people following the digging crew, tying the burlap container shut and crimping the wire basket so that it was tight around the soil ball. While it did-n't bring the highest prices, an ash was a sure sale for the nurs-eryman. Landscapers loved these trees for their toughness.

ASH is one of our most treasured native trees. There are many kinds of ash native to the United States, and many other kinds of ash native to other parts of the world, including Africa, Asia, and Europe. There are two kinds of ash that dominate American forests and landscapes: white ash and green ash. These trees are very closely related and can actually be grafted onto one anoth-er without much trouble. When European settlers first came to America they were quite taken with these trees because of their strong, straight-grained wood, which is hard to find on a tree that grows as quickly as the ash. Guitars, furniture, hardwood floors, and, most famously, baseball bats may all be produced from ash. Most of us, however, know this tree not from its wood but from its place in our boulevards, where it became very popular in the 1960s and 1970s, for an unfortunate reason.

As we learned in the last chapter, Dutch elm disease cleared streets of trees in the 1960s and 1970s. With elms quickly becoming a thing of the past, new trees had to be found to fill the niche that had once been so nicely filled by the declining elms. The new tree would be preferably something like the elm, something fast-growing and relatively strong-wooded, preferably something native. A number of different trees were planted (peo-

ple had learned their lesson and
were trying to diversify) but a few
species were planted out in greater
numbers than the others. Trees like
sugar and Norway maples were on
the rise, but the trees getting the
longest look were the ashes. These
trees just made sense. They were
native, they grew quickly (though
not as quickly as elms), they could
handle an urban environment, they
had nice color in the fall, they don't
make too much of a mess by dropping seeds, they were easy to transplant, and, best of all, they didn't
have many pests, or at least many

As the elms disappeared from
Dutch elm disease, ash took their
place admirably. (*Dave Hansen*)

pests that could threaten the tree's life. That situation changed
quite a bit in the latter part of the century.

In a small Michigan suburb just outside of Detroit, a family
decided to plant a tree. They were looking for a fast-growing
tree, not too expensive, and not too messy. They settled on a
Patmore green ash, a male that wouldn't have seeds. Unlike most
trees, ash are dioecious (There are male trees and female trees)
and so any given tree will have flowers that are one sex or the
other, but not both. Hence, a male tree will not produce seeds.
It was relatively inexpensive for such a large tree, and the man at
the garden center said that it would be easy to transplant. Only
the top of the tree was a Patmore ash. The root system was from

a seedling ash whose top had been removed. After all, with a seedling it would be impossible to tell whether the plant would be male or female, and who knew what the shape of the canopy would be? The family hired a service to come and plant the tree, which was done in good order. The father had considered moving it himself, but he strained his back at the nursery trying to lift just one side of the huge soil ball, which quickly made his decision for him. It was the only tree in the yard and would have plenty of sun. In the ground, it looked smaller than it had at the nursery, with the root system that had been obvious at the nursery now beneath the surface of the soil. The family dog, the first dog that the father and mother had adopted before the kids were born, welcomed the tree to the neighborhood with an impromptu fertilization. The kids were amused; the parents were not.

As we have seen in previous chapters, plants have been moved across the globe to satisfy humans' interests. Obviously, moving foods that we enjoy, such as peaches, to places where we can enjoy them is generally seen as a good thing. But what about insects? People have found reasons to transport insects across distances as well. As with plants, moving insects has had both positive and negative consequences. One of the most obvious benefits of moving insects over short distances is pollination. Every year, bee hives are moved from orchard to orchard across the country to ensure the fertilization of crops such as apples and blueberries. Without moving these hives, fertilization will still occur, but yield will suffer. Honey bees themselves were introduced from Europe and Asia. The first record we have of honey bees being sent to the United States was in 1621 when the

Council of the Virginia Company in London transported this insect to Virginia. Before honey bees were brought to North America, native bees pollinated plants on this continent. Native bees are mostly solitary creatures that live in small groups instead of the big hives that we're used to seeing. They aren't as efficient at pollinating many introduced crops, and they certainly don't produce huge quantities of honey. It only made sense for colonists to bring bees along with them when they came to America. But insects that have had positive impacts aren't usually what we think of when we think of introduced insects. We tend to dwell on the bad, and there are certainly plenty of introduced insects that fit that mold.

The adult gypsy moth isn't a particularly attractive creature, and what it does to trees is even worse. (*Maryland Department of Agriculture*)

Perhaps the most famous incident of moving pests occurred in the 1860s when the gypsy moth was transported from Europe to the United States to be bred with silkworms. (Yes, these species were intended to be bred with each other, something we know today to be an impossibility because most species can't interbreed.) It is rare when the introduction of an insect can be attributed to a single person, and rarer still to be able to name the address from which the insect was accidentally released, but such is the case with the gypsy moth, which was accidentally set free by a slightly absentminded Professor L. Trouvelot of 27 Myrtle Street in Medford, Massachusetts, who left his window open one day when he should have kept it shut. Trouvelot did not intend to release the gypsy moths, he intended only to try and breed them, but they got out the window nonetheless and so began our troubles. Professor Trouvelot was so upset about the escape of his gypsy moths, and so concerned about the consequences of the escape, that he tried to get the government to help him control the fugitive insects. Unfortunately, his requests were ignored.

The problem with the gypsy moth wouldn't be so extreme if it weren't for the fact that these insects attack such a wide variety of trees. They like oak, crabapples, linden, poplar, beach, birch, pine, spruce, and many others. Many of these trees are abundant across the country, and so it wasn't difficult for the moth to migrate from one food source to another. The other problem with gypsy moth is that it produces lots and lots of offspring. A female gypsy moth can easily produce over 1,000 eggs during her lifetime, which isn't an incredible number for an insect, but in most cases, the insect has natural enemies that make sure most eggs don't result in adult insects. When this moth first came to the United States, there weren't many predators that would attack it and so, when it first started out, a high percentage of the eggs actually grew into adult moths and the population grew like wildfire.

At first, the problem created by gypsy moths was purely cosmetic. These moths eat a lot, and trees were stripped to bare limbs which were significantly less attractive than trees which had leaves on them. But there wasn't that much real damage. Trees can tolerate quite a bit of leaf loss before they die, unless this leaf loss is a chronic problem, occurring year after year. More upsetting, especially to more proper individuals, was the amount of *frass* (a polite word for insect poop) that fell from trees. It was like rain when the caterpillars were feeding. The insects themselves weren't exactly a welcome sight either, and when they fell

to the road and were crushed under-
foot they made quite a mess. After
some time, trees were lost, especially
younger trees that didn't have the
nutrient stores that older trees did and
needle evergreens, which could not
quickly produce leaves to replace those
that were eaten by the caterpillars. But
as the problem got worse, even older
deciduous trees were in danger.

Charles   Valentine   Riley
(1843–95) is perhaps the most
famous American entomolo-
gist. His pioneering use of
natural enemies, pesticides,
and other methods helped
control   pests   across   the
United States and the world.
(*USDA*)

Not long after the government
declined to help Professor Trouvelot
control the initial release of the gypsy
moth, this insect became one of this
country's biggest pest problems. People
have repeatedly tried to eradicate it.
The first attempt came in 1889, which
was the first year these moths reached
what would be considered outbreak levels. The first gypsy moth
control program was instituted in Massachusetts by the world-
famous entomologist C.V. Riley. He believed that for a total cost
of $100,000 he could eradicate these insects by using a combi-
nation of insecticides and natural enemies. (The use of natural
enemies to control insects had made Riley famous in the first
place.) He used an insecticide called Paris green to attack the
moths, and had some success, but Paris green could damage
foliage and it was hard to maneuver the equipment to all the sites
where it might be needed in the forest. Besides, as we know
today, Paris green is not nearly as effective as we once thought,
and it can be quite toxic to people. Its name derives from its city
of origin and the color associated with it comes from its pur-

pose. Paris green was a paint used for shutters, and its toxicity came from the fact that one of its primary ingredients was arsenic (London purple, also based on arsenic, was a very potent insecticide as well). By 1903, the United States government had spent in excess of a million dollars to try and control this pest.

Today the gypsy moth is still a major worry, but not the worry it once was. Introduced parasites and even small ground mammals attack the gypsy moth, keeping its numbers down. Insecticide sprays have also grown to be more effective over the years and are used to keep this pest from spreading past its current range, which spreads from the Northeast south to the Carolinas and west to Wisconsin.

The ash tree grew very slowly at first, putting on only a little bit of growth its first year, but in the second and third years it started to take off and grow into the tree the family had hoped for. It could easily be seen that the tree would soon provide shade for the house and it was already providing a place for the youngest member of the family to climb, albeit only a few feet off the ground. The father decided to forego pruning the lower limbs, which would have made lawn mowing much easier, in order to provide his kids a convenient launching point for exploration into the tree's canopy.

In 1996, an insect was identified in Brooklyn, New York, infesting maples and horse chestnut trees. Ordinarily, the identification of an insect foreign to the United States is cause for concern, but not really panic. But this case was a little bit different; the Asian

longhorned beetle infesting these trees had the potential to radically change American forests. This beetle is native to Asia,
where it is not considered a major pest. Unfortunately that is not
the case here. This beetle attacks not only maple and horse chestnut, but also willow, ash, birch, poplar, and many others. As a
borer, this pest was capable of killing even healthy trees. The
Asian longhorned beetle had gypsy moth potential and worse.

Soon after the first beetles were found in Brooklyn, reports
came of beetles in Amityville and Lindenhurst in New York, and,
more disturbingly, in three separate locations in Illinois. Action
was quickly taken. Infested trees were identified and destroyed.
Destruction was a carefully prescribed chipping and burning to
avoid any escapes. Pesticides were applied to trees nearby to
avoid infestations. No trees were allowed to leave the infestation
sites, at least not without being processed. Trained scouts were
sent throughout the regions where the pest had been found to
make sure that no infested trees escaped, and word was sent
across the entire country to be on the lookout for infestations.
Signs with an odd similarity to FBI wanted posters were made
and distributed, featuring pictures of the beetle and what its
damage looked like. More reports came in. In 1999, Manhattan
and Queens had infestations; in 2002, an infestation was found in
New Jersey, and in 2003, the beetle was found in Toronto,
Ontario. New Jersey took the aggressive approach of cutting
down all possible host trees within an eighth of a mile of infestation sites.

The epicenter of the infestation in New York is an area that
conducts quite a bit of overseas shipping. It is currently thought
that the beetle came to the city sometime during the 1980s on
a pallet, or some other wooden shipping material. The presence
of this pest quickly inspired U.S. customs to require any wood
involved in shipping to be treated in such a way that would kill

the beetle. By 2008, the Asian longhorned beetle was considered eradicated from the United States, yet it reemerged that same year in Worcester, Massachusetts. No doubt this population will be quickly controlled, but for how long? Only time will tell, and even if it is eradicated this time, how long will it be before this beetle again finds its way here? Maybe in a small wooden toy brought over by a young child, or a new walking stick, or an improperly treated wooden pallet.

There has been another, less overtly threatening insect infestation building slowly over the years and attacking North American forests and landscapes. Of course, it's easy to be insidious when you're small and cute, and that's what wooly adelgids are. The balsam wooly adelgid is a very small insect, similar to an aphid, but it is covered in a waxy coating that makes it look something like a small puffball. It's only about the size of a pin head, and it loves Fraser fir and other similar firs. This insect was accidentally introduced from Europe around 1900, though it is thought to have originated in Asia, probably in Japan. Most insects like the wooly adelgid, which don't bore into trees, don't tend to be terribly deadly, but this critter is different. It can feed in the natural cracks present in a tree's bark and, as it feeds, it releases chemicals that cause the tree to grow abnormally. The vascular tissue within the tree grows in such a way that water and nutrients cannot be properly transported. The abnormal growth allows the adelgid to feed more easily. If you think this sounds something like the damage caused to grape by the grape phylloxera, then you would be right. And, as with grape phylloxera, this damage results in the plant's decline and death. Besides feeding on the trunk, this adelgid may also feed in the canopy on small twigs and buds, which leads to a slow decline from the loss of leaves. Feeding on the trunk generally leads to a much more rapid tree decline. Chemical control of this pest is possible, but

Asian longhorned beetle. This pest can attack a wide variety of trees in forests and landscapes making it extremely dangerous if it manages to escape our attempts to control it. (*Dennis Haugen, USDA, Forest Service*)

too costly for most situations. It's tactically difficult to spray an entire forest with insecticides, not to mention quite damaging to the local ecosystem. The wooly adelgid has natural enemies from other areas of the world, including certain lady beetles people have introduced into North American forests to help control this pest, but as of yet none has provided the control that we need. It would be better by far to have the situation that the Europeans do.

In its native range in Asia and Europe, the balsam wooly adelgid may be controlled by predators, but trees also do their part to help keep this pest in check, having developed a resistance to it. After all, trees can't always rely on protection from others. This resistance is based on the tree's reaction to the feeding of the insect. When the insect feeds on European silver fir, the tree's cells react by collapsing around the site of infestation. The bark formed from this collapse is inedible to the adelgid and so it must move on after causing little damage. In contrast to this the

Fraser fir is very susceptible to attacks by this insect, and we are in danger of losing significant numbers of these trees. One of the more disturbing consequences of losing Fraser firs is the loss of species related to the firs. The spruce fir moss spider is a type of tarantula found only in the southern Appalachian Mountains in moss associated with Fraser fir stands. As Fraser fir stands decline, so does the habitat filled by this endangered spider.

In its twelfth year the ash was a wonderful climbing tree because the father had left its lower branches intact. His children were getting a little old for climbing, but the kids next door would come by occasionally. All in all, the tree grew well, but its canopy seemed oddly sparse, and brown and green patches would appear in late summer that resembled a head of broccoli. To call them unsightly would be an understatement. The father sent samples to the Michigan extension service, which quickly answered his questions. The tree had ash flower gall. Nothing to worry about, the tree was fine, but they would have to live with the presence of these unsightly galls to one extent or another for the rest of the tree's life. It was much better than what the father had feared— ash yellows, which he knew was a death sentence for the tree.

WHEN ash became the new tree of choice for American boule-vards, a few insects and diseases started to be seen more frequent-ly. Most of these pests were minor, if unattractive, problems. Ash flower gall is caused by a mite that infests the flowers of male ash trees. This mite, called an eriophyid, is too small to see and isn't susceptible to many of the common pesticides used for other mites and insects. Thus, ash flower gall is usually left alone, or cut off the tree manually. While this gall can be incredibly frustrating, resulting in an ugly tree you can't do anything about, at least the tree won't die.

Ash plant bug populations also increased as ash started to dominate the boulevards. This small bug feeds on the underside of ash leaves, sucking out the sap and making the leaves appear to have

Ash flower gall. This disease, caused by microscopic mites, is extremely unattractive, but not particularly life threatening for the tree. (*Jeff Hahn*)

yellow dots. Once again, the problem is unsightly but not life threatening. And so it went, these small nuisance insects and diseases increased in frequency, but by and large the ashes were healthy—except for the occasional appearance of ash yellows. Ash yellows are a different sort of disease. Called a phytoplasma, they're related to, though not exactly the same as, bacteria, and must live in a plant's phloem to survive. Ash yellows usually shows up as a yellowing on an ash's leaves, at which point death is only a few years away. Green ash tends to be more resistant to ash yellows than white ash, and ash yellows tend occur in trees that are under stress. For this reason, as well as because of its slightly more refined looks, green ash are a somewhat more popular tree than white ash, though plenty of white ash are still planted. Ash yellows is a major disease of ash trees today, but it has never reached the extreme levels where it might be classified as a plague.

Ash provides a playground for a wide variety of pests, both insect and disease, but what is often forgotten is that these pests, in turn, provide a nice meal for other insects. There are even diseases that infect insects. In an undisturbed ecosystem, these predators, prey and diseases perform a constant balancing act. A single species will rarely dominate because of the other organisms there that keep them in check. Predatory mites will feed on eriophyid mites, and lady beetles and lacewings will feed on ash plant bugs. But when a pest is introduced that hasn't been seen before, and there isn't anything around to keep the balance, bad things happen.

There are, without question, good insects. When we think of good insects we almost invariably think of native bees pollinating trees, or native lady beetles feeding on aphids, but native insects aren't the only insects that can be good. Some of the most effective guardians of our food are introduced insects. Before the introduction of the gypsy moth in Massachusetts, another introduced insect was ruining crops in California. The cottony cushion scale was introduced to the Southwest accidentally in the late 1860s from Australia, and within ten years of its introduction it had become a serious pest that was nearly impossible to control. C. V. Riley, the same entomologist who later set out to control the gypsy moth, is best remembered for his fantastic success in controlling this pest for the California citrus industry in the 1880s. In fact, it is this success along with a few others (including some work on grape phylloxera) that led him to eventually become the USDA's head of the Division of Entomology.

Riley was a proponent of pesticides when they could be effectively used, but he quickly established that spraying pesticides to control cottony cushion scale would be only a partial solution, so he began to look for an alternative method. Knowing that this scale originated in Australia and New Zealand, and that it was rarely a problem there, he sent one of his scientists, Albert Koebele, to Australia, presuming that there the scale was kept in check by other insects native to that part of the world. Soon Koebele had identified a number of natural enemies, the most promising of which was a sort of lady beetle called the vedalia beetle. The vedalia beetle was introduced into the California citrus fields in 1888, and the cottony cushion scale was history—until the 1950s, that is, when applications of DDT reduced populations of the vedalia to such low levels that cottony cushion scale rebounded and again became a problem.

The introduction of vedalia beetle is certainly a success story and it makes bringing in predators of introduced pests seem like a great idea. But there can be drawbacks to introducing predators too. In the 1980s, a lady beetle, now called the Asian ladybeetle or multicolored Asian ladybeetle, was introduced into the United States to control aphids and protect crops, perhaps most important, pecans, where certain types of aphids can cause severe defoliation and drastically reduce the number of harvestable nuts. At first the introduction was considered a failure. The Asian ladybeetles were rarely seen and appeared to have no effect on the pest. Years later, however, the beetle reappeared with a vengeance. When it reappeared this insect was a boon to agriculture because it fed on a wide variety of aphids, including the soybean aphid, which was a newly introduced pest of soybeans from Asia in the early 2000s. Unfortunately, few people besides farmers appreciated this ladybeetle because it has some habits that we don't particularly care for, the most notable of which is its propensity for crawling into people's houses through crevices and attempting to spend the winter inside. The second and somewhat less irritating habit has to do with this insect's love of overripe fruit. Most people think of ladybugs as carnivores, and so they are for the most part, but this ladybeetle will feed on thin-skinned fruit such as raspberries and strawberries if it can get them. Remarkably, though this ladybeetle has been good for agriculture, it is thought of as a pest by most homeowners.

The ash tree grew and prospered. The ash flower gall problem was bad some years, and some years it was hardly noticeable. The family no longer thought of the tree except to admire how it shaded their house in the summer. And then one day, they were watching the news on television and saw that the city was talking about removing all the ash from a huge area close to their home.

In 2002, an insect was discovered in Michigan that was enough to turn the stomach of any tree lover. The insect was killing ash trees, and it wasn't in just a few trees—it was all over the place. No one knows exactly when the emerald ash borer arrived here, but the best guess is sometime in the 1990s, probably on some kind of packing material. If that sounds strikingly similar to the story of the Asian longhorn beetle, that's because it is. The emerald ash borer comes from Asia and attacks every type of ash on that continent to some extent and so, when it arrived in the United States, our native ash presented the insect an excellent opportunity to spread. Just as with other lethal borers, the emerald ash borer lays its eggs on trees. The eggs then hatch into larvae which burrow into the tree. As they grow, they feed on healthy vascular tissue and eventually girdle the tree, killing it. These beetles attack both healthy and damaged trees and, in some circumstances, may even be able to live on trees besides ash. In Asia, this

borer is controlled by native predators and parasites, but such controls
don't exist in the U.S., at least not
yet.

Soon after being discovered in
Michigan, the emerald ash borer
was found in Ontario and Ohio.
Everyone in and around the areas
where this borer was found was put
on alert. The newspapers, television, and radio were full of reports
telling people how to identify these
little critters and what should be

Adult emerald ash borer is attacking ash across the United States.
Because of this insect it is only a
matter of time until ash go the way
of the elms. (*Jeff Hahn*)

done if they were found. Everyone's first thought was eradication. Catch the beetle early and stop it before it becomes a problem. But it had already become a problem. This little metallic
green beetle that can take a bath in a bottle cap was being discovered in places well outside of Michigan. Indiana (2004), Maryland
(2006), Illinois (2006), and Minnesota (2009) all had populations,
probably from shipments of wood or wood packing that came
from infested areas within Michigan. Later, West Virginia,
Pennsylvania, and Quebec had confirmed sightings of the emerald ash borer, as did Wisconsin in 2008. Eradication now seems to
be a dream, but that doesn't mean that control isn't possible.

Almost as soon as the emerald ash borer was identified, efforts
were made to find out how far it had spread and how to stop it
from spreading farther. Wherever the borer was found, huge
numbers of ash were removed from that area and the surrounding land to prevent the borer from finding hosts in which to lay
its eggs. Large structures that looked something like box kites
were hung from trees. These structures held pheromones that

attracted the male beetle and let scientists know when they were present so that ash could be removed if necessary. Quarantines were quickly passed to prevent ash from being moved out of infested areas, which would allow the pest to spread farther, faster.

As the emerald ash borer spread, different methods for controlling it were investigated. Most weren't all that different than those used by C. V. Riley over one hundred years ago, except they could be put into place much quicker, thanks to the speed of modern communication. Forays were made into Asian forests to discover what insects might attack the emerald ash borer, and answers came quickly. Within only a few years, a number of different wasps were identified that parasitized the young emerald ash borer. A few of those tested seemed like excellent choices because they would attack the emerald ash borer but not harm our native beetles. In 2008, Michigan scientists released parasitic wasps into an ash forest to try and control the pest; the results will be known in a year or so.

The family saw what was happening to the ash in and around the Detroit area. They had grown attached to their ash tree and didn't want to lose its shade or its place in their landscape. It had become a part of their home. Beneath its branches was buried their first dog, and the kids had memories of climbing in its limbs. Searching for something to protect their tree from the borer, the father went to a nearby garden center and asked about chemicals that could be applied. He was quickly told by a happy employee that there was an insecticide that could be sprinkled on the ground around his tree which would enter the ground, be taken up by the tree's roots, and stop any borers from entering or damaging the tree, for a cost of about twenty dollars. It seemed like a bargain, and so the father purchased the insecticide and applied it around the tree that very night.

Borers have always been one of the hardest pests to control. Because they live inside trees they are protected from the sprays that kill most insect pests. DDT, sevin, pyrethrum, and most of the other poisons used to kill insects don't actually go into a tree, they just kill any insects that they come into contact with on the outside of the tree. Using this type of an insecticide against a borer is very difficult because the spray needs to be timed carefully so that it is present on the tree at the same time the insect is, rather than when the insect is inside the tree. This takes careful monitoring of when insects are around, and usually a lot of spraying. Such an approach would not control the emerald ash borer. Something else was needed.

Some insecticides, called systemics, enter the vascular systems of trees and other plants. They pack a nasty punch against insects; however, most systemics are considered too dangerous to be used by the average person, an exception being disyston, also known as disulfoton, which is used to protect roses from aphids. Though this pesticide is sold to the public it is one of the most dangerous pesticides to be made generally available. In the 1990s a new pesticide was being developed called imidicloprid, which belongs to a class of insecticides called neonicotinoids and behaves somewhat like nicotine, which is a nerve toxin affecting the transmission of nerve impulses. Once these nerve impulses are interfered with, the end is near. The nice thing about this poison is that, while it is systemic, it is not nearly as toxic to humans as most of the other systemics, and is effective on many types of borers. The biggest problem with this insecticide is that, as with any insecticide, it will only last for so long—a month or two at least, perhaps almost a year at best—but when it's gone, the insects can easily get into the tree and wreak havoc.

⊷

Nearby houses quickly lost their ash trees. What had once been a nicely shaded block was now only peppered with trees. But the treated trees were surviving and even looking good. In fact, fewer pests seemed to attack these trees, in part because of the insecticide being applied, and in part because there were fewer ash trees to harbor ash pests. The flower galls stayed, though. In fact, they seemed to get worse. Three years after starting to apply insecticides, the family sold their house to another family. In the transition, the first family forgot to tell the new family about keeping their ash protected. They had left a bottle of the insecticide behind for the new owners, but the new family was actually somewhat taken aback by the fact that the sellers had left behind a poison that might hurt their children, and so they got rid of it as quickly as they could through their county's hazardous waste disposal program. They believed strongly in avoiding synthetic chemicals and would never consider applying any poisons to their tree.

Two years after purchasing their house, the tree started to decline. The leaves began to drop and, when the family went out to inspect the tree, they discovered D-shaped holes on the tree's trunk, a telltale sign of the emerald ash borer. They had the tree cut down and its stump ground into the earth that fall, and replaced it with a nice, disease-resistant elm which they planted right next to the spot where the old stump lay. The elm seemed a good choice, something that would grow fast and provide shade. They reburied the dog skeleton they found under a nearby lilac.

UNCHECKED, borers attack and kill quickly. Pesticides such as imidicloprid work fine as prophylactic treatments, but they aren't a cure. As with the borers that you've already read about, the emerald ash borer eats its way into the bark and feeds on the tree's vascular tissue. Once this vascular tissue is destroyed in a ring around the stem, the tree no longer has a way to transport nutrients up and down its stem, and the tree is as good as dead.

Borers can be awful pests in and of themselves, but often they're helped by weather conditions. Just a little bit of cold damage can make it easier for a borer to enter a tree. Drought can also weaken a tree and make it more susceptible to attack. The Asian ambrosia beetle is a small insect, a little bit bigger than the size of a pinhead. It attacks a wide variety of trees and can be identified very easily by the little toothpick-like protrusions that come out of the site where it is feeding. Young trees with serious infestation end up looking like a kindergartener's craft porcupine. The beetle prefers to attack limbs or trunks that are about an arm's thickness, but they'll attack smaller or larger limbs if need be. This pest was first seen on peach trees in 1974 in South Carolina. It comes from Asia, as its name implies, and it does much less damage there than it does here. After it was first observed, it spread quickly across most of the South and demonstrated the breadth of its host range by attacking not only peach,

but also cherry, maple, plum, persimmon, oak, and many other trees.

The Asian ambrosia beetle doesn't
do its damage alone. It also carries a
fungus, called ambrosia, which actually does the digesting of the wood.
The beetle does the fungus the favor
of bringing it to its food source, the

Asian ambrosia beetle damage.
As the beetle feeds on the vascular tissue of the tree it emits long
cylinders composed of sawdust
and excrement. This is a sure
way to identify an attack by
these borers. (*Dan Horton*)

tree, and then the fungus does the
beetle the favor of allowing the beetle to eat it. It's a combination of the
fungal attack and the tunneling by
the insect that eventually kills the
tree. Since controlling the fungus is
almost impossible, we concentrate on
attempting to kill the beetle, usually by coating trees with insecticides.

This beetle will attack healthy plants, but prefers slightly
injured trees. It finds its way to trees by detecting ethylene, a
chemical released when a tree is damaged. Trees subjected to
intense cold will be more susceptible to this insect, as will trees
that are suffering because of drought. In these conditions, the
tree will be less able to protect itself. (Infested trees attempt to
push the beetle back out its hole by using sap pressure.) This beetle likes high humidity and high temperatures. Indeed, weather is
one of the factors that most predisposes a tree to an attack by this
and other insects.

In other cases, weather can do the job of killing a tree all by
itself.

Chapter Ten

FORCES OF NATURE AND ZONE PUSHERS

Wʜᴇɴ I was growing up, behind our house there was a row of lilacs planted about twenty yards from the back door. It divided our lot between the backyard, where we played, and the orchard, where we usually didn't. This row of lilacs provided a nice shady place for our dogs, Inky and Harold, to rest and for the barn cats, much too numerous to name here, to hunt and play. I grew up with the smell of lilacs and it remains one of my favorite fragrances today. The only time I can remember being without these flowers was during the five and a half years I spent in Georgia going to graduate school. Lilacs are a temperate flower and don't fare well in the southern United States. I did find another plant to serve as a rough approximation though. I worked with plants in the genus *Buddleia* while I labored as a graduate student in the hot and humid Georgia summers. These plants, alternatively called summer lilac or butterfly bush, grew quickly and had a beautiful fragrance, though it was different and much more subtle than the lilacs I was used to. When I moved from Georgia to Minnesota after graduate school to become a professor, I was excited to once again live in the land of lilacs. I also brought with me some of the summer lilacs I had collected in the South. Unfortunately, it wasn't to be. Just as lilacs couldn't handle the Georgia summers, summer lilacs couldn't handle the Minnesota winters.

Throughout this book we have seen how moving a plant from one country to another can be risky. Neither lilac nor summer lilac is native to the United States, but both are found in landscapes throughout this country. In fact, summer lilac is considered invasive in certain parts of the United States, taking over areas next to roadways and woodland borders. Lilacs aren't as invasive as summer lilacs, but we seem to plant them much more frequently—in numbers that seem to more than make up for the species's lack of invasiveness. Both the lilac and the summer lilac come from China and were moved here for one reason, their ornamental value. They are moved within our borders for exactly the same reason.

Moving a plant species from one continent to another puts the plant's health at risk. The plant will be exposed to insects, disease, and plants with which it is not familiar. They can compete with or feed on it, making its life miserable. Movement of plants within a country, especially one as large as the United States, can be risky too. I know people who plant peaches and Japanese maples in Minnesota, who bring amur maples and lilacs to Georgia, and who try crepe myrtles in Pennsylvania. These plants, introduced or not, just don't belong in the climates they're being moved to. Climatic conditions—hot and cold, humid and dry—limit the ability of plants to fill a particular patch of ground. Besides the climate, the content of the ground itself can also affect how a plant fares, as can the presence of other plants, the amount of sunlight or rainfall, and the length of the seasons. The list is endless.

Long before humans set foot on this earth, North America was covered with species of trees that aren't here today. These trees vanished for a variety of reasons, not the least of which is that the climate changed over time. If the trees in a region couldn't evolve quickly enough to remain viable, they would perish. For example, my office in St. Paul, Minnesota, was covered with ice a mere 14,000 years ago, and the land wouldn't have been able to handle even the hardiest of plants, never mind the summer lilacs that I brought with me from Georgia. Climates vary over time whether we admit it or not. But, that said, one of the most obvious changes in our climate in recent years is global warming. No matter what you believe regarding our contribution to the problem, it is difficult to dispute that the earth is warming up. Our thermometers show it, and our trees have followed suit, climbing up mountains to find conditions that better fit their needs. A study conducted by Jonathan Lenoir in Europe shows that, over the last hundred years or so, plants have shifted their preferred habitats uphill by about one hundred feet per decade, presumably to find more optimal conditions for growing. Other research shows that the size of the tropics is increasing, pushing certain species toward the earth's poles to find cooler conditions.

While natural systems favor shifting locations where particular species dwell to ameliorate the effects of climate change, humans generally ignore the changes. In our cities and landscapes we disregard the preferences of the plants we place in our landscapes, growing whatever we find attractive and can get away with. We're not likely to transplant our trees a hundred feet up a hill in order to help them grow better. In the northern United States, some people are embracing the promise of global warming, and look forward to more options for their garden. They assume that eventually they'll be able to plant things that a few years ago they could only dream about, and perhaps they're right.

By and large, trees are able to tolerate natural variations in climate in their native range, and so most weather damage to a tree planted in its home environment occurs during extreme conditions. Weather can change suddenly, brutally affecting huge expanses of land. When Hurricane Katrina hit in 2005, over 320 million trees were destroyed. This destruction affected not only the trees, but all of the creatures that lived in those trees. The loss may have contributed to global warming, as well. When an area the size of Maine is deforested and can no longer absorb any carbon dioxide, the levels of that gas will increase in the atmosphere, potentially increasing the greenhouse effect and making our world even warmer. Katrina was, of course, a one-time occurrence. But then, so are the forest fires that roar through certain states, the tornadoes that hammer the Midwest, and the other hurricanes that pound the coast.

The effects of Katrina were great, but for trees there have been far worse calamities. Mount Tambora, an active volcano in Indonesia, erupted for ten days straight beginning on August 5, 1815. The eruption buried nearby villages quickly, effectively destroying a piece of global culture. But as the largest volcanic eruption in over 1,500 years it also had other global consequences. The ash from the eruption caused the earth to cool and, in 1816, the world experienced the "year without a summer," one of the greatest global climate changes in recorded history.

Trees need their leaves in order to survive. Just as we need our mouths to take in nutrients, trees need their leaves. The leaves make sugars, which are the tree's food and the building blocks for its growth. When a late frost strikes, freezing a tree's leaves,

Trees uprooted by Hurricane Katrina in Slidell, Louisiana. (*NOAA*)

the tree loses not only the sugar already in those leaves, but also its ability to make new sugars, at least until it can send out a new flush of leaves. Trees have two ways they can cope with a late, leaf-destroying frost. The first is to avoid flushing until late in the season, after the risk of frosts is past. The second is to flush early, in order to have the capacity to flush again, just in case. If there is no late frost, trees from this second group would get a fantastic head start, and if there is a late frost, well, they're prepared for that too. What if, as in the "summer" of 1816, there was a series of late frosts that continually burned the leaves from the trees? Even trees that can produce two flushes of leaves would have suffered badly. But if the long winter of 1816 had destroyed all of the trees then there wouldn't be any trees today, and obviously there are. There are even trees alive today that were around in 1816.

As most anyone who has looked at a tree stump or log knows, trees trunks have rings, and these rings can be counted to establish how old a tree is. To the casual counter, however, 1816 is absent from the count because the ring for 1816 is too small to be noticed. Trees alive at that time did not die from the late frosts, at least not for the most part, but they did need to rely on stored food to keep themselves alive. Without new sugars trees did not grow much, if at all, that summer, and so on most trees the year 1816 is represented by only a small, insignificant ring. Some trees may not have a ring at all. Small or weak trees can die from losing their leaves in a frost, but most trees have enough sugars stored in their roots and trunk to tolerate a year without their leaves, or even two.

Of all of the catastrophes that a tree might encounter, none is as quickly fatal as fire. An ancient cause of tree death, fires are one of today's dangers that trees have had the chance to evolve with. Across the United States there are many areas prone to forest fires—some of the most notable being in Florida and California. In these locations, plants have developed seeds that can survive fires. Some trees even have seeds that are signaled to germinate by the presence of heat or even smoke, giving those species an advantage when a fire clears an area. Other trees have seeds with thick seed coats, protecting them from heat damage. Another group uses their ability to sprout back from the same stump, almost as if they were coppiced, to quickly rebound from forest fires. For these trees, fire is an effective method of getting rid of competitors. Of all of the trees introduced to the United States the one with the most interesting technique for using fire to its advantage is eucalyptus. You may know eucalyptus trees from their oil, which has a very pleasant smell. This oil is quite flammable, which happens to be very convenient for the eucalyptus tree. It is well adapted for fiery situations, and readily grows back from stumps. The trees it competes with, though, especially in its new American home, often don't have the ability to grow back as quickly, giving the eucalyptus a bit of an advantage over some of its native neighbors that aren't used to having fires ravage the forest quite as often as the eucalyptus would like.

Despite the fact that today people appreciate how trees can use fires to their benefit and how these fires are a normal part of the life of forests, they didn't always have that attitude. Once upon a time, forest fires were thought of as the end of a forest. In 1937, Franklin Roosevelt instituted a campaign to prevent our woods from burning. This is one of those ideas that seemed good at the time but which in the long run wasn't appropriate.

Forests regularly catch fire without human intervention. Leaves, twigs, and dead trees catch fire from lightning easily; add eucalyptus trees to the mix and fires become even more frequent. When forest fires are actively prevented, this debris accumulates to an unnatural level and creates a large unused store of fuel. Then, when a fire finally comes, it tends to be big and potentially uncontrollable. When people became proactive about preventing any forest fires, they promoted the accumulation of debris on the forest floor. Thus, when fires did happen, they were hotter and more out of control, and they threatened not only the forest, but also cities and homes. Today, forest managers conduct controlled burns to limit the debris. In other words, we use fires to stop fires.

In their native range, trees don't usually live in environments that aren't conducive to their growth. If an area is too warm, too cool, too wet, or too dry, a tree will lose out to other trees that are better adapted to those conditions. When humans are involved, the story is different. We enjoy creatively placing trees in environments that aren't appropriate for them. Everywhere I've lived I've encountered a certain breed of gardener I call "zone pushers." The zones, numbered 1 to 11, are climate divisions made by the USDA (United States Department of Agriculture) and reflect the minimum temperature an area will reach in a typical year. Higher numbers mean higher minimum temperatures and lower numbers mean lower minimum temperatures: Minnesota has zones ranging from 2 to 4 while Georgia has zones ranging from 6 to 8. Any tree you buy has zone(s) associated with it which can be found on the label of the plant, or, failing that, in many garden books. The zone pusher tends to think of these zones as very rough guidelines rather than gospel, an assertion that certainly seems fair considering the variable weather that can occur across regions in a given year. In some regions, the Zone number seems to reflect exceptions, not averages. Zone pushers also tend to be pH pushers (planting their trees where pH levels would dictate they shouldn't) and light pushers (planting their trees where there isn't enough light). But zone pushing is where it all begins. Cold temperatures seem to be what most people think they can most easily ignore, and why not? When a deciduous tree becomes dormant in the fall isn't it protected from those terrible cold temperatures that could kill it?

Though minimum temperatures are a big part of the difficulty with the cold, the most common damage seen on trees in cold climates actually has little to do with the temperature. It has to do with wind. During the winter, the tops of trees become dormant to avoid potential damage. Plant roots, in contrast, are not truly dormant and can take up water any time there's water present. During the winter, however, most water has become ice, and so not much water is taken up. Evergreen trees, unlike deciduous trees, have lots of leaves during the winter. These leaves make evergreens look attractive over the winter, but they also make the tree more susceptible to drying winds because they have a greater surface area from which water can evaporate. Throughout a winter, the total amount of water lost by an evergreen can be very significant, but the tree doesn't show it, remaining green until the spring. Then, in the spring, a host of needles will turn brown and fall from its limbs. Deciduous trees can suffer from the same drying out, but the water loss usually manifests in the spring as an absence of leaves on the tips of branches as opposed to the browning of the windward side of an evergreen. All this is not to say that minimum temperatures can't harm a tree, because they can and do.

When a tree is exposed to temperatures that are cold enough, the water in the tree will actually freeze. The freezing will cause water to expand, and the cells will swell and rupture. When the cells that make up the vascular tissues burst, that part of the tree that was frozen is essentially dead. Of course, the tree doesn't want its cells to freeze and it has a number of ways to stop this from occurring. The first is to get rid of excess water. Cells within the tree pump water out, into the spaces between cells where water is less likely to cause damage. Some trees also employ antifreeze. When sugar or salt is added to water, it gives the water a much lower freezing point than it normally would. The tree may use these soluble chemicals to lower the temperature at which water within cells will freeze. Finally, trees may get rid of "nucleating agents." Think of a glass. When ice starts to form, it tends to form near a crack in the glass, if one is available. Crystals (which is what ice is) like to form on edges. If the plant removes edges from its cells (via removing particles, also called nucleating agents), the water will freeze at a lower temperature. This is called supercooling. These strategies allow trees to survive the low temperatures that winters bring. And the trees that use these most effectively are the ones rated for lower zones.

Zone pushers don't take to failure easily and have a variety of tricks to help them cheat nature. Indeed, some of these tricks are necessary to bring a tender tree through the winter. Rootstocks can play an important part in cold tolerance, as they do with dwarfing and pest resistance. For example, oranges grafted onto Osage orange rootstock can tolerate lower temperatures than they could on their own. Many of the zone pushers' tricks, however, have to do with effort, not technology or genetics. Mulch and burlap are the zone pushers' best friends. Mulch piled up around the base of a plant insulates the plant's roots, the most sensitive part of the plant. Nurseries layer polyethylene, hay, and then polyethylene again over trees that are grown in containers, creating a space below this tent that stays right around freezing temperatures for the duration of the winter. Burlap is used to bundle evergreens. Like a windbreaker, burlap prevents the worst of the wind from getting to the evergreen's needles and drying them out.

ZONE PUSHING is only the beginning of what trees can suffer in the hands of the overzealous tree collector. One of the first lessons I learned in Minnesota, even before I learned how its brutal winters could make a perennial plant (such as my summer lilacs) into an annual that would only survive one season, was the lesson of pH, a measure of soil acidity. Depending on where you are, the soil may be acidic or alkaline, and the pH can make a plant thrive or fail. Some trees love the acid. Azaleas, blueberries, and rhododendrons are all acid lovers, while honey locust, bur oak, and English oak can't stand it, preferring alkaline. In their natural habitats, these plants are limited by the acidity they can tolerate. It's not that their seeds won't land in regions that are less than ideal, but the resulting, unfortunate plants won't grow after they germinate, at least not well enough to compete with other plants that are better adapted to the situation. Plants that like higher pHs (alkaline soil) are able to extract certain nutrients from the soil that other plants can't, and vice versa. Acid-loving plants cannot extract iron and manganese from the soil. Red maple and pin oak, two trees that are desired almost universally because of their beautiful form, their relatively quick growth, and their low maintenance requirements, are planted frequently across the United States. Unfortunately, these plants strongly prefer acidic, or at most slightly alkaline soils. Even more unfortunately, these were some of the first trees that I saw planted in the extremely alkaline Minnesota soils.

When a seed lands somewhere it doesn't belong, it lives, for a time. It may even live for a long time, depending on how large the tree's seed is. The bulk of a seed is made up of cotyledons, which are the primary food source for the seed until it grows leaves. These cotyledons supply the small emerging seedling with all the nutrients it needs, but like a baby as it is weaned from a bottle, the young plant needs something else to eat. As the seedling grows it must be able to obtain nutrients from the ground. If these nutrients are not forthcoming the plant will suffer and, when its cotyledon food supply is gone, the plant will die. Older trees, transplanted by humans, have the same problem as seeds, but since they are larger they have more room for food storage and so can survive longer before they expire.

Seedling red maples and pin oaks rarely survive a year in an alkaline soil; they cannot handle the lack of the nutrients and so suffer and die. Mature pin oaks and red maples can be transplanted into alkaline soils and, because they have a larger trunk than a seedling, they will have more nutrients stored; but, just like a seedling, their roots will be unable to take up all the nutrition they need. Eventually, their leaves turn yellow between the veins, they suffer dieback over the winter, and they don't leaf out as well as they should in the spring. However, the pH-pusher who decided to plant the trees can and will still claim success, at least for a few years.

Though pH and temperature can kill a tree out right, they can also bother it just enough to make it vulnerable to other problems. Such is the case with Colorado blue spruce, one of the most overplanted trees in the upper Midwest. The Colorado blue spruce is native to the central and southern Rocky Mountains, where it prefers to grow beside streams in areas of low rainfall; it is, however, a tough tree and can handle some variation. With bluish needles and an outstanding, upright form, it is without a doubt a magnificent tree, and it is understandable that everyone wants one in their front yard. Which is fine, but this spruce lives in the region that it does for a reason. It likes low rainfall, high altitudes, and a slightly warmer climate. Because of its beauty, nurseries decided to market it across the United States. It has performed well in some regions, particularly those that mimic its native range, but in other locations it ends up not looking as good as it should.

Colorado blue spruces grown in stressful environments tend to develop two diseases: rhizosphaera needle cast and cytospora canker. Rhizosphaera attacks trees when they are young, in the nursery. It causes needles to drop from the tree, making it look sparse and weak. Growing these trees can be made easier though, by spraying the plants with fungicides to prevent the spread of rhizosphaera's spores. It's a bit deceitful. The consumer buys the healthy-looking trees expecting that they are healthy when actually the trees are kept free of disease by sprays. When the tree arrives at the buyer's home and is planted, rhizosphaera often comes back, making the tree look less than spectacular. When the tree is older, rhizosphaera is less important. It is a disease that rarely kills older trees, though it can certainly make them unsightly.

As the tree gets to its tenth and twelfth year cytospora canker moves in. Cytospora's effects are deeper than rhizosphaera, and can lead to tree death. This fungus invades stressed trees (and almost all Colorado blue spruce grown outside their native range qualify as stressed), causing a sunken pit in the trunk where the disease attacks, surrounded by a raised region. This pit is where the fungus lives and is often filled with small, black, raised dots, which are fungal spores waiting to be blown to the next victim. Often, a nasty-looking ooze emanates from the infection site. At first, the symptoms are merely repulsive. Later, the fungus attacks the vascular system around the circumference of the stem, and becomes terminal. In their natural range, Colorado blue spruce grow quickly and can live hundreds of years. In an urban landscape in the middle of the upper Northwest in alkaline soils, they are lucky to last thirty.

Climates will change. Our world has experienced periods of warmth and periods of cold, and whether humans have affected global climate change or not, the earth will experience variations in its climate again and these changes will affect where plants live. But no matter how many climate changes occur, I can make one prediction with near 100 percent surety. As long as humans inhabit the earth and continue to plant trees, they will stretch the limits of these plants beyond what the trees can tolerate to get a little bit of extra color into their yards.

Chapter Eleven

LOVED TO DEATH

I once knew a twenty-five-pound cat. As most people know, twenty-five pounds is a bit too big for a domestic cat, but the woman who owned the cat loved it more than anything and fed it full servings of wonderful foods like tuna and ground beef. In the end, the overfeeding led to coronary failure. A tree can be loved to death in almost exactly the same way.

Most humans feel the need to care for other creatures: other people, animals, even trees. Have you ever inspected a tree planted in someone else's yard (not your own, of course) and in the process got your feet wet due to the amount of water applied to it? Ever seen a great mountain of mulch at the base of a tree to protect it from weeds? Ever noticed a tree that appeared to have the chickenpox from the number of black spots painted on it from a heavy pruning job? Ever wondered what the effect of this much tree love might be? We water, fertilize, stake, weed, apply pesticides to, and generally pamper the trees in our yard, providing them with much more than they would ever get in their natural setting. There is no question that many of our dogs, cats, and other pets require pampering to survive; we've bred them to depend on us. But what about trees? People have been growing trees for a long time, but have we domesticated them to such an extent that they need us for survival, or are we just pleasing ourselves at the trees' expense?

It's not uncommon for us to plant a tree when a child is born, or when someone we love dies; in fact, it's the one thing I want done to commemorate my passing. Unfortunately, those who plant trees after the loss of someone special often anthropomorphize the plant and assume it is appropriate to heap attention onto the plant as if it were the one they loved.

⋄

Once upon a time there was a tree named Gus. He was named Gus by a woman who had lost her husband, whose name was also Gus. When he passed on, she was shattered. Gus had been a great lover of the outdoors, a man who enjoyed walking through the woods, who enjoyed counting the needles on a pine tree's branch just to know how many there were. The greatest monument the woman felt she could give to her late husband was a tree. So upon his death, the woman planted one. Gus used to like to crawl into bed on cool crisp nights and snuggle with his wife. She remembered this as she planted Gus the tree, burying it deep into the earth so that the soil would come far up the stem when she filled in the hole.

The woman had purchased her tree at a nearby garden center. She selected a Bradford pear because it was a tree her husband had liked. She had seen small trees, in containers, but decided to buy one with its roots wrapped in burlap and wire. Such trees were larger, and Gus had been a big man. The tree was so

large it was hard to move, so the woman had the plant delivered
to her property. When the deliverymen arrived, they placed the
tree in the hole she had dug. She filled the hole herself, as she
felt it was her duty.

In any garden center, you'll see trees for sale in one of two packaging systems. The first is the simplest and is what most of us are
used to: container-grown trees (or, potentially, containerized
trees—which means that the trees are grown in the field and
then plopped into a container a few weeks or months before the
tree is sold). Container-grown trees are generally smaller trees
that look very nice and civilized in their containers. You can usually carry these trees by yourself or with the help of a friend. The
second packaging system is called "balled and burlapped," or
B&B for short. The name describes the package in which the
roots are contained prior to sale. A cone of soil is harvested along
with the roots of the tree using a large, heavy implement called
a tree spade. This cone of soil is then wrapped in burlap, and a
metal cage is placed over the burlap to keep the soil from falling
to pieces. No one would mistake a tree harvested B&B as something you could easily dig up yourself. These trees, along with
their balls of soil, typically weigh over two hundred pounds and
may weight significantly more depending on the size of the soil
ball. This process requires removing up to forty tons of topsoil
per acre from a nursery, and over time the loss of fertile soil
impairs its ability to grow new trees.

Big trees are attractive to buyers, and so B&B-harvested trees
are always in demand, particularly in new housing developments.
The drawback is that B&B trees don't grow well the first few

years after they are planted. Over three-quarters of a tree's root system is removed by the process of the harvest, sometimes as much as 90 percent. Compare that number with a container-grown tree, where few if any of the roots are removed. Trees harvested B&B need to reestablish their root systems before they can grow upward. In fact, these reduced root systems mean that trees harvested B&B are usually best harvested and planted in the spring or, sometimes, in the fall, when their branches are bare. Foliage greatly increases a tree's demand for water (after all, it's through pores in the leaves that most water is lost), and if the tree is harvested B&B-style while it still has leaves, the shrunken root system may not be able to keep up. In contrast, when a B&B tree is harvested at the end of winter (when the tree has no leaves), the tree can compensate for its smaller root system by simply producing fewer, smaller leaves. It takes one of these trees three years or more to grow back its root system and reach a point where it is putting on growth as fast as an undisturbed tree of the same age. One of the greatest dangers of planting a B&B tree is disturbing the few remaining roots that the tree has. When a tree is harvested B&B-style, its roots are packaged into a ball of soil, which is cinched and crimped so as not to allow the soil inside to jostle around. Jostling could cause the soil to loosen and the tree's small, fine roots to be torn, causing damage which, as you might expect, is bad news, particularly for a tree that doesn't have many roots to begin with.

Damaging a root system through ham-fisted handling is one thing; losing it through poor planting is another. Few people actually know how to plant a tree, because the process is count-er-intuitive. Who would think that planting the tree so its roots are at or even above the soil line is a good idea? People think there are many apparently good reasons to plant a tree deeply:

A tree harvested by balled and burlapped (B&B) method. The soil balls in this picture weigh between 200 and 300 pounds and are difficult to move by hand. (*Jeff Gillman*)

the roots will be deeper in the earth where they won't easily dry out; the tree will be better anchored so it won't fall over in a strong wind; there are more nutrients deeper in the earth. Unfortunately, there's a major problem with planting trees deeply that trumps all of these apparent benefits: planting deeply kills trees.

When you walk into the woods and look at the base of the trees there what do you see? Do you see roots, or do you see the trunk of the tree entering the ground? Usually, you see a tree trunk that, as you gaze down its length, flares out into roots. Roots like to be near the surface of the soil because roots need a balance of air and water. Too much water and the roots will drown. Not enough water and they'll dry out. By planting a tree deeply you're compromising the roots' water-to-air ratio. Even if the roots don't suffocate, the tree will likely perform poorly over the long haul. As roots grow, they try to achieve the perfect

Roots surrounding a stem and slowly strangling the trunk of the tree. Over time this compression will compromise the tree's ability to draw water up from its roots, or to have sugars produced in its leaves flow down to its roots. (*Jeff Gillman*)

water-to-air balance, and if a tree was planted too deeply to start with, the roots will grow upward to find this balance. As they grow upward, they lose their orientation and then, when they finally reach the balance of air and water that they're looking for, they'll start to grow outward, or potentially inward. Growing outward isn't a problem, but growing inward, toward the stem, is. When a tree's roots encounter its stem, the two don't innocently merge; instead they compress one another, competing for room—and the root is always the victor. If you have enough roots growing across a stem, the compression acts like a tourniquet, stopping the tree's vascular system from functioning and potentially killing the tree.

In its first year, the woman made sure to add lots of fertilizer, along with a special root-stimulating mix, to the hole where her pear tree was planted. She bought these products at the recommendation of one of the salesmen at the garden center, and added them to the hole before the men dropped the tree in. She wanted to make sure that Gus (the tree) wasn't hungry and that his roots spread quickly. She didn't want to make any mistakes. Afraid the soil she had pulled from the hole where Gus was to grow wasn't good enough, she purchased potting soil. She filled the hole with the potting soil, letting it crest a few inches above the surface of the ground. Then she watered the tree, setting the hose beside the hole and letting the water run slowly into it overnight.

WHEN you dig a hole for a tree, you are creating a home for it, a home that will last five, ten, even one hundred years. Thus, you want the hole to be as welcoming as possible. People often assume that the tree needs all kinds of things for its new home— fertilizer, growth stimulants, special hormones. These things are marketed as being helpful, even necessary, for a tree's survival in its new surroundings. But there is little or no evidence that these are necessary precursors to a healthy tree. All of these things are, at worst, snake oil. At best, they are questionable products that may or may not have any perceptible effect on the tree. The newly planted tree won't even be interested in taking up nutrients for about a year. There are lots of high-phosphorous fertilizers that are said to stimulate root growth. They don't. Sure, plants need phosphorus for root growth, but extra phosphorus doesn't make the roots grow any faster. In fact, too much phosphorus can inhibit root growth, and most soils have plenty of this nutrient to start with. Hormones such as vitamin B1 are also said to stimulate root growth, but success with vitamin B1 has been spotty and occurs only in the smallest of plants. Other hormones, such as Indole butyric acid, stimulate root growth on stem cuttings but may actually inhibit root growth if applied to roots that are already present. But of all the planting missteps that Gus had to endure, adding potting soil to the ground was the worst thing of all.

Soils have a "pore size," which is the measurement of spaces between particles in a soil. Water moves easily between soils with the same pore size, but not between soils with different pore sizes. So when you plant a tree and use a foreign soil, such as a potting soil you buy from the store, to fill up the hole, the hole becomes like a swimming pool; it hangs onto water, which is to say, the hole will hold a lot of water and will not quickly release it to the surrounding soil because of the differences in pore sizes between the soils. If you watered the tree when you planted it, which is what you should do, then you have placed the tree, with its reduced root system, into a small, very saturated piece of soil. Its roots won't be able to get any air, at least until the water drains away. This is a perfect situation for the onset of root disease.

❧

Gus lived through its first year though it did not flourish. The old woman watered the ground around the tree daily, hoping to keep it from dying. She saw the sparse foliage and brown top and worried about Gus's future.

Gus made it through his first winter with something less than flying colors. A few twigs had died back, but he seemed to be making it, if only barely. The next spring, he sent forth a tremendous number of blooms, which made the woman very happy. After all his flowers had fallen, the woman cut the tips off those limbs that had died the previous winter. She also pruned off some low-lying limbs that would be in her way as she mowed. She was careful to cut these limbs as close to the stem as possible, making the cuts practically flush with the trunk itself. After she had cut off all of the limbs that were giving her problems, she went to the store and bought pruning sealer and applied it methodically to every wound she had made, polka-dotting the tree with black spots. The next day she fertilized, watered, and mulched the tree again, afraid that her efforts the last year hadn't been enough.

OVERWATERING is the tree lover's bane. We know that our tree needs water, and so we give it water, and give it water, and then give it some more water. But there is a limit to how much water a tree can handle. A tree in a container needs to be watered daily because the container holds very little water; the same can't be said for a tree in a lawn. There is a lot of soil in a lawn, and in this soil there is a lot of water. Overwatering prevents roots from getting enough air, which is why trees fare poorly in areas where sprinklers go off daily to appease the water-thirsty grass. For grass, the water is great, but it is too much for the trees, and once again they suffer. But the roots of the tree aren't the only things that can suffer because of too much love and attention. So can the tree's canopy.

If overwatering is the bane of tree lovers, then pruning is the most misunderstood practice. What happens to trees in a forest if they are not pruned? The tree deals with it, dropping limbs that are sickly as need be. Most neighborhoods don't appreciate the random dropping of sickly limbs, which can be both dangerous and messy, but the truth is, while an unpruned tree probably won't appeal to our aesthetic sensibilities or be particularly safe for human traffic, it won't die either. If you walk into a forest you'll notice that there are actually very few low limbs on the tallest trees. Trees can lose their lower limbs to grazing and to crowding. Also, lower limbs don't fare well in a forest because of the overhead shade. Thus, limbs are lost and trees survive, without the aid of pruners or pruning tar.

There are a variety of ways to cut a limb off a tree. In nature they usually separate at the branch collar, which is the site of branch thickening that occurs at the location where the branch intersects the stem and extends about a quarter- to a half inch beyond the trunk of the tree. Despite nature's guidance, many people believe the best place to cut a tree's limb is flush with the trunk. To demonstrate why this is so bad it's helpful to invoke a human analogy. When dentists pull teeth, do they remove vast regions of the gums, or just pull on the teeth? The dentist will simply pull the tooth out while doing as little collateral damage as possible. That's because there is no need to damage the gums around the tooth extensively, so it's better to just remove the tooth. When you prune, think of the branch to be removed as the tooth and the branch collar and tree trunk as the gums. By damaging these surrounding tissues, you are unnecessarily extending the time that the tree needs to heal by injuring a part of the tree that doesn't need to be damaged.

Patching up limbs that have been pruned has been a standard practice for centuries. Pruning tar has been used to seal off tree wounds and prevent infections. In theory, it makes a lot of sense to seal off wounds in trees. After all, we do the same when we're injured; seal them off so that no infections can get in and to prevent blood loss. But trees don't have blood, so there's no danger of a tree bleeding out, and no one has ever shown that sealing a wound helps all that much in preventing infections. Some research has shown that using pruning sealants actually slows the creation of wood around the wound. In fact, in a pruning book from the 1920s, L. H. Bailey, one of the fathers of modern horticulture, states there is little evidence that pruning tar works at all. Little new evidence has materialized since its publication. So why do we still use the sealer? An excellent question and one that I'm sure someone at your local garden center would love to address.

Gus survived his first few years. As the woman grew older she had less and less energy to water the tree. She was lucky to get out once a week, never mind once a day. She fertilized once a year, at most. And Gus grew stronger and larger than he had before. He had suffered a slow start, but now he was looking better, putting on growth and developing a full canopy. The woman still got out sometimes to sit in the shade and enjoy a book. She would read old novels and even gardening books. The tree didn't bloom as heavily one year and so the woman made a plan for the next year, a plan she had read about in one of her husband's old gardening books by Jerry Baker. She would beat the tree with a baseball bat to encourage it to flower. Around the tree's base she applied mulch, lots of it, so that if she couldn't apply it again next year there would be enough to suffice. She stacked it up around the trunk, making what looked like a small volcano. Almost a foot deep at the stem, the woman knew that the mulch would keep the ground moist and stop weeds from stealing Gus's water and fertilizer, which she applied with gusto.

A mulch volcano placed against a tree's stem. This mulch can encourage disease to form on the tree's trunk and stem girdling roots. (*Jeff Gillman*)

TREES are constantly under some sort of stress: too much water, too little water, too hot, too cold. Strangely, stress inspires trees to reproduce, a principle that was once used to make trees flower more heavily. Beating a tree with a baseball bat and damaging its vascular system would indeed cause a tree to put forth more flowers. Unfortunately, the ensuing year, in which the tree might develop more flowers, might also be the tree's last because the battering would damage the tree's vascular system.

Mulch is so good, and yet so bad. While defending the ground around the tree from weedy invaders, wood mulch turns slowly into soil. After all, plants that have decomposed make up a large percentage of most healthy soils. As the mulch turns to soil it has the same effect on the roots as planting a tree too deeply. The soil level rises as the mulch decomposes. As the soil level rises, the water table rises slightly, and in turn the roots need to grow higher to find the proper water-to-air balance. The roots find their way up into the mulch and grow across the stem, potentially strangling the tree to death. Mulch also tends to create a humid environment, an environment that fungi and bacteria very much enjoy, so there is the potential for diseases to infiltrate a tree from the mulch that surrounds its trunk.

Properly applied mulch is only a few inches think and is arranged in a doughnut around the tree, leaving the shallowest portion immediately around the stem. Just like trees in the forest that die and decay, thus feeding other trees in the vicinity,

mulch provides nutrients as it breaks
down. But people tend to want more
rapid responses and so are quick to
add fertilizers to their trees to inspire
them to grow.

If you make a habit of fertilizing
your lawn once in a while, then there
is really little need to fertilize your
trees very much. The grass will not be
able to absorb all of the nutrients that
are applied to it, and the nutrients
that move down below the surface of
the grass's roots, will be collected and
used by the trees. Once a tree grows

Trees in landscapes are constantly under stress. This tree was destined to live a short life from the
moment it was planted.
(*Gary Johnson*)

large enough its extensive root system will be able to mine the
nutrients it needs from the soil and will need your help even less.
This isn't to say that fertilizers used on trees are useless, because
they're not, but it is safe to say that they're overused and that
more of the fertilizer runs off the property where it is applied
than ever finds its way into the tree. In fact, many trees don't use
more than 20 percent of the fertilizer applied to them.

As the next year came, the woman found herself in need of help.
A stroke had robbed her of her ability to walk, and she had to
rely on her family to care for the pear tree. Her daughter moved
into the house and, per the woman's request, paid lots of attention to Gus. After a year, the daughter decided to marry the man
she had been dating for three years, in part because she knew her
mother would not be around much longer and she very much

wanted her to be a part of the wedding. Afterward, the husband moved in and they decided to add on to the house. Originally, they had planned to build the extra room right where Gus was standing, but the mother would not have it, so they built the addition about ten feet from Gus's trunk. Gus was getting to be a large tree now, his stem almost eight inches in diameter. The old woman still found time at least once a week to sit in Gus's cool shade and remember her Gus and how he would have enjoyed sitting with her.

IF trees could become scared, then one of the most frightening sights would surely be that of bulldozers and wheelbarrows gathered on a site where people intend to build. Trees don't like construction because construction changes things, and trees don't adapt well to change. Even if a tree is purposely "saved" by not being run over with a cement mixer, any construction close to that tree will potentially sever a portion of the root system, causing the part of the canopy connected to those roots to grow more slowly. Construction also compresses the earth, which is hard on older trees with established root systems. As bulldozers, trucks, and carts scurry across a site, it forces the earth underneath to compact and changes the water relations of the area. Over time some spots will be moist that had been dry and others will be dry that had been moist. Construction can also change the overall level of the soil. When a new foundation is dug for a house or an addition, a certain amount of soil is displaced, and the contractors have to lay it somewhere, often around a nearby tree. Putting soil near a tree isn't the best of ideas because, as with mulch, it raises the level of the soil and changes

the level of the water table, raising
it up to a point where the roots
will, once again, need to grow
upward to maintain the proper
water-to-air balance. There is no
surer way to damage a tree during
construction than by raising the
level of the earth next to that tree,
but there are more insidious ways.

Stressed trees are hotter than
unstressed trees, and by photographing a forest from above using
an infrared camera it is easy to tell
where in the forest trees are struggling. This heat derives from the
fact that trees under stress do not

Half of the canopy of this tree is
fine, and half is defoliated because
nearby construction has compromised half its root system.
(Jeff Gillman)

transpire as much as unstressed trees. Transpiration is the evapo-
ration of water from the leaves and, like sweat does for humans,
it cools trees down. Transpiration occurs in urban forests as well,
and by photographing parks with infrared cameras you can
establish which portions of that park are under the most stress.
Unsurprisingly, trees nearest to where people walk and play are
usually most stressed.

Gary Johnson, a professor of urban forestry at the University
of Minnesota, was doing this sort of photography in Minnehaha
Falls Park, the site of recent construction in Minneapolis, when
he noticed something unexpected. At this park a staging ground
for concerts and plays was moved from one section of the park
to another. When this occurred there were a significant number
of trees under stress, as expected, around the area where the new
staging ground was built and where the old staging ground was

torn out. However, there was also a great deal of stress outside of these two areas, more stress than there should have been. When Gary went and examined the apparently highly stressed trees that weren't near the staging area, he found that they followed a broad path that led from the parking lot to the new staging area. This is a popular park and over the course of a weekend it isn't uncommon to have 5,000 or so people visit, potentially walking over this area. The planners had decided not to install a direct walkway from the parking lot to the new staging area, and instead tried to encourage patrons to use other paths that guided people on a more roundabout journey to get to the area. Naturally the people rebelled and followed the most direct route regardless of whether walkways had been installed or not.

To satisfy himself that the damage was actually being done by the rebellious visitors, he took measurements to test soil compaction. He used a penetrometer, a device that quantifies ground compaction by measuring the force required to jam the contraption (which is basically a pointy, stainless-steel stick with a pressure meter on the top) into the ground. Gary took a reading, which showed that the ground was only slightly compacted. A year later, he took another reading and found that it took four times as much force to drive the penetrometer into the ground as it had the previous year.

What was the effect of the compacted earth on these trees? After all, they were sixty- and seventy-foot tall bur oaks, and apart from giving off excess heat they seemed very healthy. At first glance, they appeared the same as their neighbors. When you got close to these trees, however, there were some differences. Bullet galls, small growths that are shaped like old-fashioned musket balls, were detected on the oaks' stems in higher quantities than on other, healthy oaks in the vicinity. Bullet galls are caused by a wasp that lays its eggs onto the stems of oaks. Besides injecting an egg, the wasp's sting also causes the stem tissue to grow oddly, and it is on this oddly growing tissue that the young wasp feeds. Larval wasps then grow up in these galls until they emerge from them as adults. These galls don't typically kill trees, but they do cause a tree stress, which is bad for a tree that is already suffering due to compacted soil. At the time this book is being written, there is some concern that the two-lined chestnut borer will move in and cause more damage. This borer is a native insect that infests and eventually kills oaks by feeding on their vascular system. It is a common pest of bur oak and is especially aggressive when it finds a stressed tree it can take advantage of.

To try and fix the problem the earth around the oaks was decompacted by vertical mulching, a practice accomplished by digging vertical trenches and filling them with mulch to alleviate compression. Now we'll never know for sure whether the stress of all of the people marching over the tree's roots would have eventually killed them, but I have a feeling the park managers and patrons are comfortable with their ignorance. Sometimes it's better not to know what might have been.

⌒

The woman passed away on a cold February day with her daughter by her side and her pear tree visible outside the window. Gus was starting to fade a little. The notoriously weak branches of the Bradford pear were falling off from the weight of the winter snow, tearing bark and wood from the trunk as they dropped to the ground, useless but still connected. One branch fell and hit the house, causing some damage to a gutter. The daughter and her husband, now the owners of the house,

called four different services about tree pruning, but the quotes
they received from each seemed much too high for the simple
job of pruning a tree. The fifth company they contacted gave
them a much lower quote, and so they set a date for the pruning
that spring.

When the company arrived, they found Gus rather large and
rough looking. Because of the shape it was in, with the limbs on
the side of the house where the addition had been built a few
years before showing obvious signs of decline, they recommended that the tree be removed. The daughter immediately rejected
that idea. She knew that her recently deceased mother wouldn't
have stood for it. Gus deserved better, and she told the men as
much. So the men from the tree care company compromised
and recommended at least thinning the branches so the wind
couldn't blow them around and cause wind damage to the tree.
They also recommended taking off the top of the tree so that if
a part of it fell due to high wind or heavy snow, the house would
be safe. The couple agreed and the men got to work in the late
spring.

There were three men in the crew. Two of them climbed the
tree while the third, the manager, instructed them where to cut.
He told them to cut out the interior branches of the tree. When
they got to the top of the tree they cut the branches in a uniform height, giving the tree a sort of flat-top. After they were
done, the men received their pay and left. The couple went out
to look at the tree and, while they weren't entirely pleased with
it aesthetically, they weren't displeased either. The new look certainly was orderly. They were especially pleased that the tree had
been thinned to remove the leaves that would catch the wind
and put stress on branches causing them to break. By removing
all of the leaves from the interior of the canopy and leaving those

on the outside alone the tree appeared to still have a full canopy; the heavy thinning could be seen only if you were to stand underneath the tree and look up. All in all, the couple was happy with their choice to prune the tree.

In most towns and cities, numerous companies are willing and able to prune trees. These companies vary widely, however, in their ability. As a rule, certified arborists do the best work and know the most about your trees, while handymen do the worst job and know the least. Unsurprisingly, handymen tend to charge significantly less for pruning a tree than certified arborists. The saying "you get what you pay for" applies well here. Inexperienced tree pruners usually know the basics of tree pruning—they know not to make flush cuts, they know essentially what a canopy needs to be healthy. Some cuts, however, seem like good ideas when they are not and can deceive inexperienced or untrained pruners. It seems like a good idea to thin out the small, leafy branches on the interior of a tree's canopy, while allowing those leaves on the outside to remain. After all, the canopy still looks good, and the inside of the canopy has been thinned out, thereby reducing surface area on which the wind can push and seemingly putting less stress on the branches when the wind blows. This is not really the case, however. When this type of pruning is done, you are left with branches that resemble lion's tails, with a bare branch that terminates in a clump of leaves. This configuration puts an incredible amount of stress on the tree limb when the wind blows because only the tip of the limb is moved by the wind, rather than the whole thing, causing the branch to whip in the wind more than it normally

A large topped tree. This type of pruning encourages the growth of many new branches around the removed limbs. (*Dave Hansen*)

would. While this is bad enough, topping is worse. Topping is an old practice, once thought to be beneficial, but now known to be quite damaging.

Topping is the practice of cutting back the top of a tree so that it's essentially flat. This technique is detrimental because it destroys the tree's leader. Most trees have a branch that is taller than the rest and provides direction for the tree's growth. The tip of this branch produces more auxin—the plant hormone we looked at in chapter 5 (it promotes tree growth)—than any other limb on the tree. Because of this leader, most tree canopies develop a conical configuration. The loss of the leader is a problem for the tree, and it will need to choose a new branch to serve this function. Normally, this transition is not difficult. The tree simply selects the next tallest limb and allows that limb to take over the role of the leader, but for a tree where the entire canopy has been cut flat, the decision is difficult. With all of its branches at a single height, the tree struggles to figure out which branch should be the new leader. For a short time, all the branches act like a leader, growing straight upward. This upward growth is terrible for the tree because it creates very narrow branch angles between tree stems. The angle between two branches which are next to each other and growing straight up will be small. As these branches grow, they will get thicker and eventually press together at their bases forming areas of bark inclusion—where bark is pressed between growing limbs. These inclusions represent very weak branch-to-stem attachments and will be prone to breakage in the future.

ᐅ

Gus produced many new shoots after the pruning. The top of his canopy looked like a messy, spiked haircut with no single limb outcompeting the others to be the new leader. About the time the pruning was being done, something else was happening, too. The couple noticed brown lesions and curling leaves on the tree. They didn't worry, at first, but when it got worse they decided to send a sample to their county agent, an employee of the state university who could identify plant diseases. They sent the sample off and received a response in a couple of weeks: the tree had fire blight. The note went on to say that they should avoid pruning during the late spring. Unfortunately that suggestion was now too late.

Fire blight had entered the cuts made by the men when they pruned Gus's canopy. A few brown leaves were the least of the tree's worries. These leaves could grow back, so long as the roots were healthy. Over the course of the year the bacteria responsible for fire blight sat essentially dormant in the tree's limbs, and then, the following year, the couple noticed that many of Gus's limbs were not producing leaves and that, later in the spring, many of the trees leaves were dying off. By the end of the year the couple had to hire the tree care men again, this time to remove Gus. Though it broke the couple's hearts, they knew keeping him alive was a losing battle and thought it best to put him out of his misery.

ONE of the worst diseases to affect pears, apples, hawthorns, and cotoneasters, along with a host of other plants, is fire blight. This bacterial disease invades a plant's xylem, the tissue that transports water and nutrients up to the leaves, and can cut a tree down to size in no time flat. Most fire blight you see in the landscape isn't a big deal, just a few brown leaves here and there on your favorite pear tree. But when fire blight enters a tree's vascular system, often through a cut on a stem or branch, the tree is in serious danger. The bacteria can live in xylem for up to a year without causing symptoms. However, once it starts to spread, the tree is in trouble. Fire blight will destroy the xylem, producing a thick ooze, which is a collection of sugars that are toxic to the plant. The simple presence of ooze signals that a plant is terribly infested and usually foretells the loss of limbs, at least, and the entire tree, at worst. Fire blight is a common end to many poorly cared-for pears and apples, and even some that are well cared for.

Throughout Gus's life the tender care he received was actually killing him, and then, having survived this thoughtful care he suddenly developed a disease where, if thoughtful care wouldn't have saved him, it might at least have slowed his demise. But that's the way it is with trees. So often we think we know what we're doing with these majestic towers, only to be taught that we're don't know half as much as we think we do.

Chapter Twelve

THE FUTURE
OF THE TREES

Most state universities have diagnostic clinics for plants where, for a small fee, you can have your plant's maladies examined. These clinics can usually identify diseases, insects, or even nutrient problems, and they often receive the most interesting samples. I've never worked in one, but when I first started working in Minnesota, my office was across the hall from a diagnostics clinic and I often got to see the more interesting samples that came across their desks. I'll never forget the time that someone sent in a picture of a small tree wanting to know what type of borers had made the remarkably large holes that were evident on the tree's trunk. Larger than the exit holes of most borers, these holes had the people in the clinic stumped for a while. Eventually though, someone noticed that the "borers" had a diameter strangely similar to that of a bullet from a .45-caliber handgun.

Usually these clinics are able to diagnose the problems they see, but every once in a while there's something that's so strange or off-the-wall that they can't figure it out. My favorite clinic story comes from Warren Copes, a plant pathologist with the USDA in Mississippi. Before Warren began working with the USDA, he worked at a diagnostics lab in Athens, Georgia. All kinds of samples crossed his desk, and one of the most common diagnoses he made was dog blight. Dog blight isn't exactly a disease, but more of what you might call a cultural problem. When dogs urinate, they release a lot of nitrogen onto whatever it is they're urinating on. Oftentimes this nitrogen is so concentrated that it's too much for grass, shrubs, or even trees to handle, and so it kills a portion of the plant. This phenomenon is commonly known as dog blight.

One day, Warren received a sample from an arborvitae whose leaves were turning brown. The sample that the tree owner sent in was, as is usually the case, just a few leaves. The sample was set into a dish of nutrient solution that would grow any fungi or bacteria that might be on the leaf including any common diseases that the arborvitae may have developed. Once those organisms had grown for a short time they could be identified. After a few days of sitting in an incubator, the dish was examined and it was determined to be disease free. When a sample comes in to be tested, besides just putting it on a dish to grow plant diseases, some notes are also taken, and the notes from this plant said the leaves had smelled like ammonia. As a rule, if there is an ammonia smell to a sample, and if no disease is evident, then the cause of the problem is likely to be dog blight, and so that is the diagnosis that Warren sent to the arborvitae owner. The arborvitae owner did not, however, agree with Warren's diagnosis, and she gave him a call to let him know it. Unbeknownst to him, the sample the woman had sent had come from the very top of the arborvitae, which was about six feet off the ground, and unless there was a dog that rivaled Clifford in size, or a dog who had a remarkable ability to project urine vertically in a water-gun like stream, dog blight was an impossibility. They were at a standstill. Warren was not going to go out to the property from which the arborvitae came in search of an explanation, and the woman was not going to send more samples and pay more money; she considered Warren's diagnosis ridiculous. Then, a few days later, the woman called Warren back to apologize. On the side of her house where the arborvitae sat was a raised porch, where her grandchildren liked to play, and one of the games that they played was pee into the bush, which just so happened to be the very arborvitae in question.

So far I've spent a great deal of time and paper looking at how trees have died historically, and how trees are dying now. But what about the future? What lies in wait for the trees? Things aren't going to change. There will be disease, insects, and the saw, but there are also new dangers for trees. One of these emerging dangers can be summarized as follows: we are, like the little boy who thoughtlessly peed from his grandmother's porch, poisoning our trees. We are expelling our wastes into the environment and the trees are left to deal with it. Over time, our air has suffered from the consequences of our careless dumping of chemicals into the atmosphere. The first indication that pollution was taking a toll on our trees was acid rain. Though acid rain is thought of as a problem of the late 1900s, it was actually identified in the late 1800s by Robert Angus Smith in his book *Air and Rain: The Beginnings of a Chemical Climatology*, first published in 1872. Smith was a chemist who recognized that our airborne pollutants cause rain to become acidic and be potentially damaging to trees. One of his most famous comments is about the effect of industry-dumped chemicals in the air in Manchester, England, where he noted that "rain instantly reddens test paper in Manchester, and where most soot is found there is most acidity. It is easy to account for the absence of plants." Despite his findings, however, the existence of acid rain was considered an academic concern until the 1960s, when it became clear that the wastes being expelled from industries, particularly those whose power was based on coal, were having a pronounced effect on our trees.

The term *acid rain* paints a mental image in my head that is more fiction than fact. When I hear the term, I think of buildings melting under a drizzle of pure hydrochloric acid and of open sores sizzling from acid reactions on skin (it's possible I read too many comics as a kid). That isn't the way acid rain works at all. As a matter of fact, acid rain is rarely strongly acidic and doesn't appear any different than normal rain. Acid rain is usually caused by nitrogen oxides and sulfur dioxide, which are released into the air when coal and oil are burned. In the air, these chemicals react with other constituents of the air and with water to form very dilute solutions of nitric and sulfuric acid. These solutions are so dilute that you could drink them and nothing bad would happen. Nothing would happen to most plants, either, unless they were constantly watered with the stuff and the soil wasn't well buffered.

Buffering is a term that refers to the ability of something, in this case soil, to resist change. In the case of acid rain, a well-buffered soil will not react very strongly to the rain and so the rain won't affect the ground much at all. A poorly buffered soil, on the other hand, will not be able to resist the acidity of the acid rain and so the soil's pH will slowly become more acidic. Despite common perceptions, acid rain that falls on a tree is not likely to be overly damaging to the tree's canopy, unless the tree happens to reside high on a mountaintop. In high, mountainous regions trees are often affected by acid rain to a greater degree than at lower elevations. That is because at higher elevations, trees tend to sit in fog and mist for longer periods of time. If this mist is composed of acidic water droplets, then the plant's leaves are under a great deal of acidic stress, which may over time lead to the tree's death. Trees in lower elevations don't experience much damage to their leaves due to acid rain, besides a small reduction in photosynthesis. Acid rain is most damaging to the soil in which the tree's roots sit, particularly if that soil is poorly buffered. In poorly buffered soil, acid rain prevents certain nutrients from being as available to the plant and causes other elements, which the plant doesn't need much of, to become overabundant. One of these elements is aluminum. When soil pH becomes acidic, aluminum becomes very available to the plant, causing aluminum toxicity and perhaps even death. At the very least the tree becomes stressed and more susceptible to attack from other pests.

While acid rain still exists to varying degrees across much of the world, legislative efforts, and the efforts of industry, have greatly reduced the amounts of nitrogen oxides and sulfur dioxide that are released into the air. Methods are in place across much of the world to remove these polluting chemicals from the exhaust of smokestacks before they reach the atmosphere. The story of acid rain is largely, though not completely, a success story. Because we are still burning fossil fuels, we are still releasing some nitrogen oxides and sulfur dioxide into the atmosphere. These chemicals still produce local acid effects.

Besides acid rain, industrial waste also produces ozone, a chemical created when nitrogen oxides react with the oxygen naturally in the air. An ozone molecule is made up of three oxygen atoms, and in the world of oxygen atoms, two's company and three's a crowd. The extra oxygen atom wants to relinquish its relationship with the other two and react with something else, and that something else is often a tree. One of the most visible manifestations of ozone is as one of the most damaging constituents of smog. And, although we usually think of smog as something that quickly dissipates, ozone does not go away quickly, taking about three weeks to disappear from the troposphere (the lowest part of our atmosphere, where we live). As the most abundant and injurious air pollutant out there in the atmosphere right now, ozone poses a significant problem to trees wherever it is present.

In cities the threat of air poisoning trees is very real. Ozone levels as low as a hundred parts per billion or lower can cause visible damage to sensitive trees. Ozone poisoning usually manifests on trees that are growing in urban environments where there are a lot of cars, or a lot of industry. Just as some trees are more susceptible to some insects, so some trees are more susceptible to ozone poisoning. Willow, green and white ash, and loblolly pine are all very likely to show ozone damage, while sugar maples, ginkgos, and arborvitaes tend to be tolerant of this pollutant. Ozone poisoning rarely kills a tree outright, however. Instead, this pollutant weakens a tree and makes it more susceptible to insects and disease, and less adaptable to environmental changes such as soil compaction or overwatering.

In the San Bernardino Mountains in Southern California, major constituents of the forests are the ponderosa and Jeffrey pines, both of which are particularly sensitive to ozone. During

The San Bernardino Mountains. (*The Wildlands Conservancy*)

the 1960s and 1970s, this area was bathed in heavy, ozone-rich smogs that came into the mountains from Los Angeles. Paul Miller, a researcher who has been studying these trees, found that from 1974 through 1988 the ponderosa and Jeffrey pines had put forth less growth than other, similar species that were competing with them for space and sunlight. Because of this the ponderosa and Jeffrey pines had given way to white pine, sugar fir, and black oak in areas where heavy ozone damage had occurred. The damaged upper canopies of the original pines were no longer able to prevent light from reaching the forest floor, and so the growth of other trees was encouraged. Besides the evolution of the forest toward more ozone-tolerant species, a new danger was emerging in the area: fire. The low branches and thin bark that are characteristic of the ozone-resistant species are more flammable and make forests susceptible to forest fires, which is likely one of the reasons that the wildfires in the San Bernardino Mountains in 2003 reached the magnitude which they did, destroying over 92,000 acres of land.

Levels of ozone are dropping in today's world. The drop is slow, but it's real. As the smog problems in L.A. lift, slowly, so does the burden of ozone in the San Bernardo Mountains. Until we wean ourselves off fossil fuels, we will continue to create at least some pollution of the environment from using these resources. But it seems the biggest effect that burning fossil fuels have on our trees has yet to be realized.

Besides nitrogen oxides, sulfur dioxide, and ozone there is another chemical that may hurt our trees. A chemical whose release is rarely controlled, and which is fundamentally associated with the burning of fossil fuels: carbon dioxide. Fossil fuels come from long-dead animals and plants that have slowly, over millions of years, turned into a mass of carbon-containing compounds, the most common of which are oil and coal. The ancient plants that make up these compounds originally acquired their carbon the same way modern plants do, from the carbon dioxide in the air. When these materials are burned for power they are forced to return to the air what they once consumed and carbon dioxide is released. There has been an incredible amount of press on the possibility that all the carbon dioxide we're releasing could make Earth a warmer planet, but most of this press has ignored discussing the effects of the excess carbon dioxide on plants. People assume that because carbon dioxide is needed for photosynthesis, an increase in carbon dioxide in the atmosphere will be a good thing for plants and their growth, but that isn't necessarily the case. For some trees, higher levels of carbon dioxide could have much more serious consequences.

Plants consume carbon dioxide and react it with water, using sunlight as an energy source to make sugars. Oxygen is released as a waste product. Over time, plants have evolved different ways to take up the carbon dioxide. Some have developed what is known as a C_3 pathway, others a C_4 pathway, and still others what is known as a CAM pathway, which, for our purposes, is very similar to a C_4 pathway. The difference between a C_3 pathway and a C_4 pathway is that a C_4 pathway is more efficient at taking up carbon dioxide, and so the pores on plant leaves that do this job don't need to stay open as long. Thus, the plant is able to conserve water, since not as much will be lost through its open pores. However, this system also limits the amount of carbon dioxide that can be taken up. In other words, extra carbon dioxide in the air is beneficial to C_4-pathway plants, but only up to a point. Once that point is crossed, the plants simply cannot absorb any more. In contrast to C_4 plants, C_3 plants are less efficient at acquiring carbon dioxide but they have a higher threshold for taking it up. Most trees are C_3 plants; however, there are quite a few important plants that use C_4 or CAM pathways, such as corn, sorghum, millet, sugar cane, and orchids. Due to the differences between C_3 and C_4 plants, as the level of carbon dioxide increases in our world, certain plants will benefit more than others (with C_3 plants usually coming out on top).

Besides the obvious differences between plants with different carbon pathways to take up carbon dioxide there is also the issue of efficiency. Of all of the plants that have been tested at high levels of carbon dioxide, trees tend to respond the best, but some trees simply utilize extra carbon dioxide more efficiently than others do and so, as carbon dioxide levels increase, some trees will be able to take advantage of this while others won't. Exactly which species do best isn't quite clear just yet, but, as with ozone, you can bet that some trees will fare well, and some will fail. And this is ignoring the effects that carbon dioxide increases may have on the warmth of our globe. There are so many variables to take into account when considering the effects of carbon dioxide on trees it's difficult to know where to even start. Suffice it to say that by enriching the world with carbon dioxide we will be changing which trees grow and thrive in different areas.

OUR forests, orchards, and yards bear things of great beauty. Sometimes we treat trees as they deserve to be treated, with reverence and caring, and other times we use them as a crop. Sometimes we harm them despite our best efforts, and sometimes, despite out best efforts, trees manage to survive. There is a theory about parasites and prey that says the perfect parasite takes what it needs from its prey without causing it much harm. As a matter of course, the parasite may eventually develop symbioses with its host, where both parties benefit from the relationship. Such is the relationship between mycorrhiza, a group of fungi that associates with the roots of various plants including trees. The mycorrhiza takes carbohydrates from trees while offering the trees nutrients that they collect as they grow under the surface of the soil.

When a parasite is first introduced to its prey the interaction is awkward. Often, the parasite kills the prey or leaves it so damaged that the prey is unable to recover. If the two survive their first meeting, however, the parasite and prey evolve together. After all, if you can't keep from killing all of your prey, eventually you won't have anything to eat. I like to think that humans' relationship with trees is similar. The forests have suffered extensive damage at our hands as this country grew. Today human-caused changes have altered what the forests contain and what they can grow. Even the density of our forests has changed. But we can learn. As we evolve with the forests, perhaps we can become a more effective parasite than we have been in the past, taking what we need without pushing these ecosystems beyond what they can tolerate.

In front of my parents' house in Pughtown, Pennsylvania, sit two silver maples, planted in 1977, the first year we lived there. My wife, children, and I usually travel to Pennsylvania once or twice a year to visit. My older daughter has reached the age where she wants to see what her father did when he was younger, and so we enjoy walking through the woods, investigating the old stream, and climbing the old trees, just as I did when I was young. I can even point out places where I changed something twenty or even thirty years ago, such as a pushed-over pile of stones, or a carved place on tree bark. It surprised me recently when I discovered that my daughter's favorite place isn't the woods, or the stream, or the small copse of spruce, where you can hide. Instead, it's the place right in front of the house where two silver maples sit. One of the maples was improperly cared for when it was young and now it has a low branching pattern. The branches have narrow angles between them, and are a little bit too close together. Also, the branching is a little too high to make for a really good climbing tree. Still, it's not bad. Unfortunately, a silver maple is a notoriously short-lived tree, and I don't expect it to last much longer, relatively speaking of course. But for now, my daughter thinks it is just about the best climbing tree she has ever seen. Together, we've chased down a cicada killer wasp, with prey in tow, and had the chance to see huge wheelbugs attack small caterpillars lower in its branches. We've also named caterpillars and seen galls on its leaves, not to mention the games of make-believe with princesses and mermaids. (I'm still partial to games of army and cowboys and Indians, myself.) As time goes by, I'm hoping the tree will continue to keep a place in her heart, and in her sister's too. There's something eternal about the love that we hold for trees, something that shouldn't fade away but should be treasured always.

BIBLIOGRAPHY

New works stand on the shoulders of old. This book incorporates first-hand experience and interviews with those experts mentioned in the acknowledgments. In addition, I consulted a variety of valuable written sources for information, including the following.

Chapter 2. A Charmed Life

Johnson, W. C., and T. Webb. 1989. The role of blue jays (*Cyanocitta cristata* L.) in the postglacial dispersal of fagaceous trees in eastern North America. *Journal of Biogeography* 16: 561–571.

Meyer-Berthaud, B., S. E. Scheckler, and J. Wendt. 1999. Archaeopteris is the earliest known modern tree. *Nature* 398: 700–701.

Stein, W. E., F. Mannolini, L. VanAller Hernick, E. Landing, and C. M. Berry. 2007. Giant cladoxylopsid trees resolve the enigma of the Earth's earliest forest stumps at Gilboa. *Nature* 446: 904–907.

Chapter 3. Forests Old and New

Beckwith, J. R. III, and K. D. Coder. 2005. Growing quality wood from CRP pines. Bugwood Network, Extension Forest Resources, University of Georgia.

Carter, D. R., and E. J. Jokela. 2002. Florida's renewable forest resources. University of Florida IFAS Extension CIR1433.

Daniel, T. W., J. A. Helms, and F. S. Baker. 1979. *Principles of Silviculture* (2nd ed.). New York: McGraw-Hill.

DeLoach, T., and D. Bales. 2009. Southern pine beetle. Mississippi State University Extension Service. http://msucares.com/forestry/titleiii/articles/spb.html.

Doran, F. S., C. W. Dangerfield, F. W. Cubbage, J. E. Johnson, J. W. Pease, L. A. Johnson, and G. M. Hopper. 1998. Tree crops for marginal farmland: Loblolly pine. Virginia Cooperative Extension Service Publication number 446–604.

Kellogg, G. A. 1914. Redwood, king of Humboldt county. *Davis Commercial Encyclopedia of the Pacific Southwest.*

van Mantgem, P. J., N. L. Stephenson, J. C. Byrne, L. D. Daniels, J. F. Franklin, P. Z. Fulé, M. E. Harmon, A. J. Larson, J. M. Smith, A. H. Taylor, and T. T. Veblen. 2009. Widespread increase of tree mortality rates in the western United States. *Science* 323 (5913): 521–524.

Monroe, Martha. 2000. Trees in your life. IFAES. FOR 81.

Rackham, Oliver. 1980. *Ancient Woodland: Its History, Vegetation and Uses in England.* London: Edward Arnold.

Schmidtling, R. C. 2001. Southern pine seed sources. U.S. Forest Service General Technical Report SRS-44.

Schultz, R. L. 1997. Loblolly pine: The ecology and culture of loblolly pine. USDA Agricultural Handbook 713.

Smith, W. B., P. D. Miles, J. S. Vissage, and S. A. Pugh. 2002. Forest Resources of the United States, 2002. USDA North Central Research Station, St. Paul, Minn.

Zamora, J. H. 2006. The "Grandfather" of Oaklands redwoods. *SFGate* August 14, 2006.

Chapter 4. Stowaway Plants

Correll, D. S. 1962. *The Potato and Its Wild Relatives.* Renner, Texas: Texas Research Foundation.

Kloeppel, B. D., and M. D. Abrams. 1995. Ecophysiological attributes of the native *Acer saccharum* and the exotic *Acer platanoides* in urban oak forests in Pennsylvania, USA. *Tree Physiology* 15: 739–746.

Radcliffe, E. B., and D. W Ragsdale. 1993. Insect pests of potato in Minnesota and North Dakota. In: Bissonnette, H. L., D. Preston and H. A. Lamey. Eds. Potato Production and Pest Management in North Dakota and Minnesota. Extension Bulletin 26 AG–BU–6109-S.

Spongberg, S. A. 1990. *A Reunion of Trees.* Cambridge, Mass.: Harvard University Press.

Webb, S. L., T. H. Pendergast, M. E. Dwyer. 2001. Response of native and exotic maple seedling banks to removal of the exotic, invasive Norway maple (*Acer platanoides*). *Journal of the Torrey Botanical Society* 128(2): 141–149.

Wilson, C. L., and C. L. Graham, eds. 1983. *Exotic Plant Pests and North American Agriculture*. New York: Academic Press.

Wyckoff, P. H., and S. L. Webb. 1996. Understory influence of the invasive Norway maple (*Acer platanoides*). *Bulletin of the Torrey Botanical Club* 123(3): 197–205.

Chapter 5. Apples and Age

Hokanson, S. C., and C. E. Finn. 2000. Strawberry cultivar use in North America. *HortTechnology* 10 (1): 94–106.

Hughes, J. d'A. 1989. Strawberry June yellows—A review. *Plant Pathology* 38: 146–160.

Knight, T. A. 1795. Observations on the grafting of trees. *Philosophical Transactions of the Royal Society of London*. 290–295.

Loehle, Craig. 1987. Tree life history strategies. *Canadian Journal of Forest Research* 18: 209–222.

Munne-Bosch, S. 2007. Aging in perennials. *Critical Reviews in Plant Sciences* 26: 123–138.

Owen, James. 2008. Oldest living tree found in Sweden. *National Geographic News,* April 14, 2008.

Pyle K. 1999. Legacy of an apple seed. *American Forests,* Spring 1999.

Rom, R. C., and R. F. Carlson. 1987. *Rootstocks for Fruit Crops*. New York: Wiley-Interscience.

Shabtai Cohen, S., A. Naor, J. Bennink, A. Grava, and M. Tyree. 2007. Hydraulic resistance components of mature apple trees on rootstocks of different vigours. *Journal of Experimental Botany* 58 (15–16): 4213–4224.

Shigo, A. L. 1996. Reading the tree's log. *Tree Care Industry* 7 (6).

Smart, C. M. 1994. Gene expression during leaf senescence. *New Phytologist* 125: 27–58.

Westwood, M. N. 1993. *Temperate-Zone Pomology* (3rd ed.). Portland, Ore.: Timber Press.

Chapter 6. The Short Lives of Peach Trees
Blackburne-Maze, P. 2003. *Fruit: An Illustrated History.* Toronto: Firefly Books.

Gould, H. P. 1918. *Peach-Growing.* New York: Macmillan.

Haire, B. 2004. Georgia peach crop best in decade. *Southeast Farm Press.* June 7, 2004.

Horton, D., and D. Johnson, eds. 2008. *Southeastern Peach Growers Handbook.* University of Georgia College of Agricultural and Environmental Sciences. http://www.ent.uga.edu/peach/peachhbk/toc.htm

Okie, W. R., C. C. Reilly, and A. P. Nyczepir. 1985. Peach tree short life-effects of pathogens and cultural practices on tree physiology. *Acta Horticulturae* 173:503-512.

Ordish, G. 1987. *The Great Wine Blight.* London: Sidgwick and Jackson.

Rieger, M. 2006. *Introduction to Fruit Crops.* Binghamton, N.Y.: Haworth Food & Agricultural Products Press.

Smil, V. 2001. *Enriching the Earth.* Cambridge, Mass.: MIT Press.

Chapter 7. Foreign Invasion
Carter, N., and J. Llewellyn. 2002. Sharka (plum pox virus) of stone fruit and ornamental Prunus species. Ministry of Agriculture Food and Rural Affairs, Ontario. Factsheet Agdex# 634.

Gordon, F. C., and D. A. Potter. 1985. Efficiency of Japanese beetle (Coleoptera: Scarabaeidae) traps in reducing defoliation of plants in the urban landscape and effect on larval density in turf. *Journal of Economic Entomology* 78: 774–778.

———. 1986. Japanese beetle (Coleoptera: Scarabaeidae) traps: Evaluation of single and multiple arrangements for reducing defoliation of plants in urban landscape. *Journal of Economic Entomology* 79: 1381–1384.

Green, A. 2000. Crawford sounds alarm on screwworm found in Florida. Florida Department of Agriculture and Consumer Services. Department press release 10-05-2000.

Metcalf, C. L., and W. P. Flint. 1939. *Destructive and Useful Insects* (2nd ed.). New York: McGraw-Hill.

Metcalf, C. L., W. P. Flint, and R. L. Metcalf. 1951. *Destructive and Useful Insects* (3rd ed.). New York: McGraw-Hill.

Smith, J. W. 1998. Boll weevil eradication: Area-wide pest management. *Annals of the Entomological Society of America* 91 (3): 239–247.

Stefferud, A., ed. 1953. *Insects: The Yearbook in Agriculture.* Washington, D.C.: United States Government Printing Office.

Chapter 8. The Plague of the Elms

Gibbs, J. N. 1978. Intercontinental epidemiology of Dutch elm disease. *Annual Review of Phytopathology* 16: 287–307.

Griffin, L., G. Griffin, and J. R. Elkins. 2009. American Chestnut Cooperators Foundation. www.ppws.vt.edu/griffin/accf.html

Mittempergher, L., and A. Santini. 2004. The history of elm breeding. *Invest. Agrar: Sist. Recur. For.* 13 (1): 161–177.

Osborn, A. 1945. *Shrubs and Trees for the Garden.* London: Ward, Lock & Co.

Peace, T. R. 1932. The Dutch elm disease. *Forestry* 6: 125–142.

Chapter 9. Unstoppable Insects

Antonelli, A. 2004. Balsam wooly adelgid: A pest of true fir species. Washington State University Extension EB1456.

Baker, J. R. 1994. Cottony cushion scale. North Carolina State Cooperative Extension. ENT/ort-51 July 1994.

Chenot, A. 1996. The history of gypsy moth control in the United States. *Midwest Biological Control News Online* 3 (8).

DeBach, P., and B. R. Bartlett. 1951. Effects of insecticides on biological control of insect pests of citrus. *Journal of Economic Entomology* 44: 372–383.

Doane, C. C., and M. L. McManus, eds. 1981. The gypsy moth: Research toward integrated pest management. U.S. Department of Agriculture, Forest Service, Science and Education Agency. Technical Bulletin 1584.

Hoebeke, E. R. 2003. Invasive insects as major pests in the United States. *Encyclopedia of Pest Management* 1 (1), 1–4.

Liu, H., T. R. Petrice, L. S. Bauer, R. A. Haack, R. Gao, and T. Zhao. 2003. Exploration for emerald ash borer in China. http://emeraldashborer.info/files/EABchina.pdf

McCullough, D. G., A. Agius, D. Cappaert, T. Poland, D. Miller, and L. Bauer. Accessed 2009. Host range and host preference of the emerald ash borer. USDA Forest Service and Michigan Department of Agriculture. www.emeraldashborer.info/files/hostrangeDEB.pdf

Skerl, K. L. 1997. Spider conservation in the United States. University of Michigan Endangered Species Update. http://www.umich.edu/~esupdate/library/97.03-04/skerl.html.

Chapter 10. Forces of Nature and Zone Pushers

Chambers, J. Q., J. I. Fisher, H. Zeng, E. L. Chapman, D. B. Baker, and G. C. Hurtt. 2007. Hurricane Katrina's carbon footprint on U.S. Gulf Coast forests. *Science* 318 (5853): 1107.

Keeley, J. E., 1987. Role of fire in seed germination of woody taxa in California chaparral. *Ecology* 68 (2): 1987.

Lenoir, J., J. C. Gegout, P. A. Marquet, P. de Ruffray, and H. Brisse. 2008. A significant upward shift in plant species optimum elevation during the 20th century. *Science* 320 (5884): 1768–1771.

Lovett, R. A. 2007. Climate change pushing tropics farther, faster. *National Geographic News,* December 3, 2007.

Chapter 11. Loved to Death

Bailey, L. H. 1927. *The Pruning Manual.* New York: Macmillan.

Baker, J. 1972. *Plants Are Like People.* New York: Pocket Books.

Gillman, J. H. 2006. *The Truth About Garden Remedies: What Works, What Doesn't and Why*. Portland, Ore.: Timber Press.

Chapter 12. The Future of the Trees

Ainsworth, E. A., and S. P. Long. 2005. What have we learned from 15 years of free-air CO_2 enrichment (FACE)? A meta-analytic review of the responses of photosynthesis, canopy properties and plant production to rising CO_2. *New Phytologist* 165: 351–372.

Appleton, B., J. Koci, R. Harris, K. Sevebeck, D. Alleman, and L. Swanson. 2000. Trees for problem landscape sites—air pollution. Virginia Cooperative Extension, Publication 430-022.

Environmental Protection Agency (EPA) Acid Rain Website: http://www.epa.gov/acidrain/effects/ accessed 2008.

Gorham, E. 1982. Robert Angus Smith, F.R.S., and "Chemical Climatology." *Notes and Records of the Royal Society of London* 36 (2): 267–272.

Miller, P. R. 1996. Extent of ozone injury to trees in the western United States. USDA Forest Service General Technical Report PSW-GTR-155.

Smith, R. A. 2007 (first printing 1872). *Air and Rain: The Beginnings of a Chemical Climatology*. Whitefish, Mont.: Kessinger Publishing.

INDEX

ACKNOWLEDGEMENTS

I first decided to put this book together at the encouragement of Bruce H. Franklin, to whom I owe a great debt. I was in Bogota during the summer of 2007 and received an odd e-mail from Bruce, whom I did not know at the time, asking whether I knew anyone who might be interested in writing a book on how trees die, or if I might be interested in writing the book myself. When I first read his e-mail I thought it was a silly idea. After all, trees die all the time for all kinds of reasons, and who would really want to read about that? But after a day or two of listing all the ways trees can die, I changed my mind, and this book is the result.

When I first started to put pen to paper, words came easily. After all, I'm a horticulturalist who works with trees and other plants, and so I thought I knew a little bit about how trees are grown and how they die. But I soon discovered that my knowledge was more limited than I thought, and so I enlisted the help of other experts who knew much more than I did about various topics. My old friend Dan Horton from the University of Georgia was very helpful with fruit crops, particularly with peaches, and I hope he is not too disappointed at my choice of death for the peach in chapter 6. Emily Hoover, Jim Luby, and Stan Hokanson were also extremely helpful when it came to fruit, particularly apples. Aging in trees is a difficult topic, but I found conversations with Alan Smith, Neil Anderson, and David Zlesak to be very enlightening, and they pointed me to all the right places. Christian Thill, the most potato-obsessed person I have ever met, provided insight into potatoes whose interesting

story certainly belongs in a book on trees. Jeff Hahn provided
insight regarding insects, both big and small.

I have little experience with managed forests besides my random
wanderings, mostly through southern pine groves, and so
conversations with Carl Voigt and Anthony D'Amato were
invaluable. Gary Johnson, a constant help to me in all of my
work, was also extremely helpful over the course of this book's
creation, offering critiques and advice at key times. I also thank
Warren Copes, whose story about dog blight graces the last
chapter of this book, and Janna Beckerman who pointed me in
Warren's direction.

My wife, Suzanne, put up with all of my random musings as
did Chad Giblin, my partner in tree research who is also the person
who keeps things running while I'm off writing. Buddy
Hardee gave the material a good reading and provided strategic
input. Thomas Boyle provided key comments and edits to this
book. I wrote the words, but Tom made the words work. This
book would be much less without his efforts. Thanks also to
Noreen O'Connor-Abel who provided useful editing and comments
throughout this text.